Nonequilibrium Statistical Mechanics *of* Heterogeneous Fluid Systems

Andrei G. Bashkirov, D.Sc.
Leading Scientist
Laboratory of the Earth Revolution
United Institute of the Physics of the Earth
Russian Academy of Sciences
Moscow

CRC Press
Boca Raton London Tokyo

Library of Congress Cataloging-in-Publication Data

Bashkirov, Andrei G., 1940–
Nonequilibrium statistical mechanics of heterogeneous fluid systems / Andrei G. Bashkirov.
p. cm.
Includes bibliographical references and index.
ISBN 0-8493-2860-8
1. Statistical mechanics. 2. Hydrodynamics.
3. Nonequilibrium thermodynamics. I. Title.
QC174.8.B37 1994
530.4′27—dc20 94-23470
CIP

Direct all inquiries to CRC Press, Inc., 2000 Corporate Blvd. N.W., Boca Raton, FL 33431.

International Standard Book Number 0-8493-2860-8
Library of Congress Card Number 94-23470
Printed in the United States of America 1 2 3 4 5 6 7 8 9 0
Printed on acid-free paper

This book is dedicated to the memory of Dmitrii N. Zubarev

Andrei G. Bashkirov, D.Sc., is Leading Scientist at the Laboratory of the Earth Evolution at the Institute of the Physics of the Earth, Russian Academy of Sciences, Moscow. Born in 1940 in Moscow, he finished secondary school while being awarded the Gold Medal. In 1963, he received his university degree from the Physical Department of the Moscow State University and continued his postgraduate studies in statistical physics at the Mathematical Institute of Russian Academy of Sciences under the supervision of Prof. D. N. Zubarev.

Dr. Bashkirov has worked in various fields of nonequilibrium statistical physics and kinetic gas theory. Originally, he was interested in the statistical theory of the Brownian motion and the nucleation considered as consequences of long-range fluctuations in a fluid. His main interests at the present time are the kinetic gas theory of the shock wave and the theory of dispersed systems, among them the systems of gravitating particles.

Dr. Bashkirov was the scientific editor of Russian editions of various English language textbooks and monographs in statistical physics and kinetic theory. During ten years he was a member of the Soviet Editorial Board of the journal, *Fluid Dynamics – Soviet Research.*

Preface

In our age of scientific and technical revolution a development of technique and technology is tightly interwoven with a development of fundamental investigations. In particular, successes in aviation and space, chemical engineering, and electronics demand the construction of advanced new approaches describing transport phenomena in intrinsically inhomogeneous media with interfaces.

First and foremost there are phenomena at interphase boundaries: liquid–gas, liquid–liquid, liquid–solid, and gas–solid. In the traditional approaches interphase phenomena are jump changes in physical parameters, i.e., in the form of boundary conditions for equations describing properties of bulk phases. However, in general, surface properties possess their own dynamics, thus, their description can not be reduced to the usual boundary conditions for bulk phase equations. In this sense it is possible to say that the new field of science, that is, physics and mechanics of nonequilibrium surface phenomena, is emerging.

A wide variety of disperse systems are gaining acceptance in different technological processes including suspensions, emulsions, and fixed and fluidized beds of catalysts, etc. Each such system is an assembly of two or more bulk phases, divided by interphase surfaces. One of the bulk phases is usually a fluid and another is solid or liquid particles. Every particle, as a macroscopic formation, possesses a finite surface in contrast to molecules that are customarily considered as point particles. This difference results in the main distinguishing feature of dispersed systems, that is, interactions in these systems are determined by flows of matter, momentum, and energy across interphase surfaces but not interaction potential, as it takes place in molecular systems.

When particles are sufficiently small, fluctuations of the flow across the particle surfaces can appear to be essential as it occurs, for example, in the cases of Brownian motion of particles in a fluid or the growth of a liquid phase embryo in a supersaturated vapor.

Hence there is an intrinsic connection between the theory of surface transport processes and the dispersed system theory, which is reflected in this book.

The general nonequilibrium statistical approach, due to Zubarev, is presented briefly in Chapter 1. This approach is widely used in this book as a starting point in the studies of surface systems.

Chapter 2 is devoted to construction of an application of this approach to a statistical mechanical description of transport processes on a dividing surface between two immiscible fluids when singular (excess) densities of mass, and/or momentum, and/or energy are presented. The surface hydrodynamics equations are derived here for the excess densities. When there are no excess surface densities, these equations transform into boundary conditions for parameters of the bulk phases on the boundary surface.

In Chapter 3 the shock wave in a gas is considered an interphase boundary of the gas–gas type. A new parameter for the gasdynamic discontinuity is found, i.e., the surface tension coefficient, and the relevant surface (capillary) pressure appears in the modified Rankine–Hugoniot jump conditions. The concrete gasdynamic problem of shock wave instability is resolved here with the use of these new jump conditions.

In Chapters 4 and 5 fluctuations of flows of mass or momentum across the surfaces of sufficiently small liquid or solid particles, respectively, are discussed. These fluctuations are responsible for the growth of a liquid phase embryo in a supersaturated vapor, or the Brownian motion of a solid particle in a fluid.

Finally, Chapter 6 discusses a generalization of the one-particle Brownian motion theory to many-particle situations, where the interparticle hydrodynamic interactions are essential.

I have tried to keep the mathematical formalism as simple as possible. I assume, however, that the reader is familiar with the elements of applied mathematics, basic laws of mechanics and thermodynamics, and the elements of statistical mechanics.

It is a pleasure to thank the numerous friends an colleagues who have contributed their comments, thoughts, and ideas over the years. In particular, I have had the good fortune to have as a teacher and long–time senior friend, Prof. Dmitrii N. Zubarev, to whose memory this book is dedicated.

I am especially grateful to my colleague Andrei V. Vityazev. His help and promotion have been of crucial importance. I extend thanks to the staff of the United Institute of Physics of the Earth for helpful cooperation in preparing of this manuscript.

Contents

Chapter 1

Zubarev Method of Nonequilibrium Statistical Operator

1.1 Contracted Description of a Statistical System

The main mission of nonequilibrium statistical mechanics is to derive equations describing irreversible processes, starting from the Liouville equation for the distribution function in the phase space of the system. It is known, that this equation is reversible relative to time inversion, in other words, say this equation is invariant relative to simultaneous changes of the time direction and the signs of all particle velocities (or momenta).

To obtain equations describing irreversible processes, it is necessary to proceed to a less detailed description of the system. It is the Bogoliubov concept of the hierarchy of relaxation times at the evolution of a nonequilibrium statistical system that is of first importance in a mathematical formulation of this less detailed or contracted description.

The Bogoliubov concept [1] in its original application to moderately dense gases is as follows: to describe an N–particle system at any arbitrary initial state we are to assign a full N–particle distribution function or a rich variety of many–particle time–dependent distribution functions. However, after a short "time of synchronization" occurs, time dependence of distribution functions becomes fully determined by the evolution of a one–particle distribution function. As a result, after some characteristic time, we have a simplification or a contraction of the description and as a consequence it is possible to describe the further evolution of the N–particle system with the use of fewer parameters. In the case of a dilute gas we use the one–particle distribution function. Thus, there is a kinetic stage of evolution.

Later, after a next characteristic time, the next contraction of the system description occurs, using fewer parameters. For a dilute gas there are some first momenta of the one–particle distribution function, for example, the density, the temperature, and the mean velocity of gas as functions of the time and space coordinates. So we have a hydrodynamic stage of evolution.

For moderately dense gases, the time of synchronization of many–particle distribution functions is of order of the duration of the intermolecular interaction, i.e. the collision time, and the setting of the hydrodynamic stage of evolution occuring at the mean free path time.

This picture becomes more complicated in general many–particle systems.

For example, there is no apparent kinetic stage of evolution described by the one–particle distribution function in liquids. In any case each successive stage of evolution corresponds to a greater degree of a chaotization of the system and to a less detailed description ofits evolution. At a later stage we have fewer representative variables to describe system motion, i.e. a smaller number of degrees of freedom.

On the other hand, while we prescribe a reduced set of parameters, each related to a particular collective dynamic variable, we thereby prescribe the time scale wherein the adopted description of the system is valid. It is significant that, in this time scale, the distribution function depends on time only via the prescribed parameters. Thus at this stage of the evolution it is possible to restrict the problem to a solution of the Liouville equation in a particular form, namely as a functional $\rho(t)$ of the reduced set of parameters.

1.2 Quasiequilibrium Distribution Function

The distribution function $\rho(t)$ in the form of a functional of a reduced set of parameters may be found by applying the relevant boundary (or initial) condition to the Liouville equation. It is defined using a quasiequilibrium distribution function $\rho_q(t)$.

Here we discuss the construction of a quasiequilibrium distribution for a classical (non-quantum) system, wherein a nonequilibrium process is described by a reduced set of functions of the dynamic variables $\{P_m\}$. These functions may, in general, depend on the spatial coordinate $\mathbf{x}$. If we agree upon inclusion of this dependence into index m, it can then take both discrete and continuous values. Therefore, a summation over m may include an integration over a continuous subset of values m.

A quasiequilibrium distribution function for the reduced description is

$$\rho_q(t) = \exp\{-\sigma(t)\} \tag{2.1}$$

where

$$\sigma(t) = \ln \mathrm{Sp}[\exp(-F_m(t)P_m)] + F_m(t)P_m \tag{2.2}$$

Here $F_m(t)$ are parameters thermodynamically conjugated to the variables P_m; it is supposed the summation (or integration) with respect to the recurrent indexes m, and Sp[...] is a multi-dimensional integral over the full phase space of the system.

The distribution function $\rho_q(t)$ realizes a conditional extremum of the information entropy functional:

$$S(t) = -\mathrm{Sp}[\rho_q(t) \ln \rho_q(t)] \tag{2.3}$$

under the conditions

$$< P_m >_q^t = < P_m >^t \tag{2.4}$$

and the normalization condition

$$< 1 >_q^t = 1 \tag{2.5}$$

where

$$< ... >_q^t \equiv \mathrm{Sp}[...\rho_q(t)] \tag{2.6}$$

$$< ... >^t \equiv \mathrm{Sp}[...\rho(t)] \tag{2.7}$$

These conditions imply that at any instant t the parameters $F_m(t)$ are defined by equations (2.4) in such a manner that the quasiequilibrium distribution function $\rho_q(t)$ would give the mean values $< P_m >_q^t$ of all variables of the contracted description coinciding with their observable values $< P_m >_q^t$.

We also consider two the most characteristic particular cases of the quasiequilibrium distribution (2.1).

An equilibrium state of a many-component system is the state of the most full contracted description. In this case the main integrals of motion, namely an energy $\mathcal{H}$, a momentum $\mathbf{P}$, and particle numbers N_α (α is the index of the component) are chosen as the variables $\{P_m\}$ of the contracted description. So the distribution (2.1) becomes the grand canonical equilibrium Gibbs distribution

$$\rho_0 = Q_0^{-1} \exp\{-F_1\mathcal{H} - F_2\mathbf{P} - \sum_\alpha F_{3\alpha}N_\alpha\} \tag{2.8}$$

$$Q_0 = \mathrm{Sp}[\exp\{-F_1\mathcal{H} - F_2\mathbf{P} - \sum_\alpha F_{3\alpha}N_\alpha\}] \tag{2.9}$$

As parameter $\{F_m\}$ we have

$$F_1 = \beta, \qquad F_2 = -\beta\mathbf{v}, \qquad F_{3\alpha} = -\beta(\mu_\alpha - m_\alpha v^2/2) \tag{2.10}$$

where $\beta = 1/kT$, T is the absolute temperature of the system, k is the Boltzmann constant, $\mathbf{v}$ is the velocity of a motion of the system as a whole, μ is the chemical potential, m is the molecular mass.

The grand canonical distribution (2.8) gives the possibility for determining all thermodynamic properties of the systems and fluctuations of the energy, the momentum, and the particle number of a system.

If there are no fluctuations of momentum and particle number in a one–component system, we may confine a contracted description of the system to one variable, namely, the energy $\mathcal{H}$. In this case we obtain the canonical Gibbs distribution function

$$\rho_C = Q_C^{-1}\exp\{-\beta\mathcal{H}\} \tag{2.11}$$

$$Q_C = \mathrm{Sp}[\exp\{-\beta\mathcal{H}\} \tag{2.12}$$

This distribution also gives the possibility to determine all thermodynamic properties of the system and energy fluctuations.

Another characteristic case is a weakly nonequilibrium inhomogeneous state of a many-component fluid. In this case densities of energy $H(\mathbf{x})$, momentum $\mathbf{p}(\mathbf{x})$, and particle numbers $n_\alpha(\mathbf{x})$, in a three–dimensional coordinate space are taken as variables $\{P_m\}$.

In accordance with such a choice of $\{P_m\}$ the parameters $\{F_m\}$ are the time– and space–dependent generalizations of those defined by equation (2.10), in other words, we introduce fields of temperature $T(\mathbf{x},t)$, velocity $\mathbf{v}(\mathbf{x},t)$, and chemical potentials $\mu_\alpha(\mathbf{x},t)$:

$$\begin{aligned} F_1(\mathbf{x},t) &= \beta(\mathbf{x},t), \qquad F_2(\mathbf{x},t) = -\beta(\mathbf{x},t)\mathbf{v}(\mathbf{x},t) \\ F_{3\alpha}(\mathbf{x},t) &= -\beta(\mathbf{x},t)(\mu_\alpha(\mathbf{x},t) - m_\alpha v^2(\mathbf{x},t)/2) \end{aligned} \tag{2.13}$$

The distribution function $\rho_q(t)$ in this case takes the form

$$\begin{aligned} \rho_l(t) = Q_l^{-1}\exp\{-\int d\mathbf{x}\big[&\beta(\mathbf{x},t)H(\mathbf{x}) - \beta(\mathbf{x},t)\mathbf{v}(\mathbf{x},t)\cdot\mathbf{p}(\mathbf{x}) \\ &-\beta(\mathbf{x},t)\sum_\alpha\big(\mu_\alpha(\mathbf{x},t) - m_\alpha v^2(\mathbf{x},t)/2\big)n_\alpha(\mathbf{x})\big]\} \end{aligned} \tag{2.14}$$

$$Q_l(t) = \mathrm{Sp}[\exp\{-\int d\mathbf{x}\big[\beta(\mathbf{x},t)H(\mathbf{x}) - \beta(\mathbf{x},t)\mathbf{v}(\mathbf{x},t)\cdot\mathbf{p}(\mathbf{x}) \\ -\beta(\mathbf{x},t)\sum_\alpha\big(\mu_\alpha(\mathbf{x},t) - m_\alpha v^2(\mathbf{x},t)/2\big)n_\alpha(\mathbf{x})\big]\} \tag{2.15}$$

This distribution function is the direct generalization of the grand canonical Gibbs distribution; equation (2.8), describing the local equilibrium state of an inhomogeneous fluid, permiting the determination of all thermodynamic properties of the system.

Historically, the concept of local equilibrium was first proposed by Mori in 1958 [2]. A detailed discussion of this concept is, for example, in textbooks [3, 4].

1.3 Nonequilibrium Distribution Function

It is necessary to stress that the quasiequilibrium distribution function, equation (2.1), and its particular form, the equation (2.12) do not satisfy the Liouville equation for a distribution function

$$\frac{\partial\rho(t)}{\partial t} + i\mathcal{L}\rho(t) = 0 \tag{3.1}$$

Here $\mathcal{L}$ is the Liouville operator; its action on any function A of dynamic variables is defined as

$$\mathcal{L}A = -i\{A, \mathcal{H}\} \tag{3.2}$$

where $\mathcal{H}$ is the Hamiltonian of the system, and $\{\ ,\ \}$ is the Poisson bracket operator.

The shortcoming of the quasiequilibrium distribution implies that it cannot be used for description of nonequilibrium processes in the system. Thus we shall look for a solution for the Liouville equation (3.1), corresponding to the same level of a contracted description as the quasiequilibrium distribution function (2.1).

We will follow the method proposed by Zubarev [4] for handling this problem. In accordance with this method, the quasiequilibrium distribution function is used as a boundary (or initial) condition to the Liouville equation. Moreover, this boundary condition is formulated with the use of an infinitesimal source in the right–hand–side of the Liouville equation. So, it takes the form

$$\frac{\partial\rho(t)}{\partial t} + i\mathcal{L}\rho(t) = -\varepsilon(\rho(t) - \rho_q(t)) \tag{3.3}$$

This source permits us to choose solutions for the Liouville equation describing irreversible processes associated with an increase in entropy (when $\varepsilon > 0$). In the infinite past these solutions correspond to the quasiequilibrium distribution

$$\rho(t)|_{-\infty} = \rho_q(t)|_{-\infty} \tag{3.4}$$

In evaluating nonequilibrium properties of the system with this solution of this kind and approaching to the thermodynamic limit ($V \to \infty$, $N \to \infty$, N/V =const; V and N are the full volume and the particle number of the system), it is necessary to approach the limit $\varepsilon \to +0$. Then the main feature of the Liouville equation solution is conserved, i.e. the time derivative of the average of any dynamic value evaluated using this solution is equal to the average of the time derivative of this dynamic value.

We note here that in equilibrium statistical mechanics, the use of infinitesimal perturbations distorting any symmetry of a system is the starting point of the method of quasiaverages by Bogoliubov [5]. This method was proved to be very useful in theories of superfluidity, superconductivity, phase transitions, ferromagnetism, etc.

The same procedure is well known in mathematical physics. By way of illustration, the theory of the Helmholtz equation is recalled. This equation for a amplitude of harmonic oscillation $u(x)$ takes the form

$$\Delta u + k^2 u + f(x) = 0, \qquad k^2 > 0 \tag{3.5}$$

To select a retarded solution complying with Zommerfeld radiation condition, an infinitesimal positive source ε is introduced, so the equation becomes

$$\Delta u_\varepsilon + (k^2 + i\varepsilon)u + f(x) = 0 \tag{3.6}$$

The solution to this equation is unique and corresponds at $\varepsilon \to +0$ to the retarded (causal) solution to equation (3.5) and at $\varepsilon \to -0$ to the advanced (anticausal) solution to equation (3.5).

The modified Liouville equation (3.3) determines the nonequilibrium distribution function for a contracted description. Due to the boundary condition, equation (3.4), the time symmetry of the Liouville equation is broken and the retarded solution $\rho(t)$ is defined as a functional of the function $\rho_q(t)$ and hence as a functional of thermodynamic parameters $F_m(t)$, that, in turn, are determined by equation (2.4).

We rewrite equation (3.3) as

$$\frac{\partial}{\partial t}\Delta\rho(t) + \{\varepsilon + i\mathcal{L}\}\Delta\rho(t) = \rho_q(t)\dot{\sigma}(t) \tag{3.7}$$

where

$$\Delta\rho(t) = \rho(t) - \rho_q(t), \quad \text{and} \quad \dot{\sigma}(t) = (\frac{\partial}{\partial t} + i\mathcal{L})\sigma(t) \tag{3.8}$$

Equation (3.7) may be represented in an integral form. If $\Delta\rho(t)$ is assigned at an initial time t_0, we have

$$\Delta\rho(t) = U_\varepsilon(t,t_0)\Delta\rho(t_0) + \int_{t_0}^{t} dt_1 U_\varepsilon(t,t_1)\rho_q(t_1)\dot{\sigma}(t_1) \tag{3.9}$$

where

$$U_\varepsilon(t,t_0) = e^{-\varepsilon(t-t_0)} \exp\left\{-i\int_{t_0}^{t} dt\mathcal{L}\right\} = e^{-\varepsilon(t-t_0)}U(t,t_0) \tag{3.10}$$

is the evolution operator for equation (3.7).

In equation (3.9) approaching the limit $t_0 \to -\infty$ and taking into account the boundary condition, equation (3.4), we get

$$\Delta\rho(t) = \int_{\infty}^{t} dt_1 U_\varepsilon(t,t_1)\rho_q(t_1)\dot{\sigma}(t_1) \tag{3.11}$$

Explicitly performing the differentiation, $\partial/\partial t + i\mathcal{L}$, in equation (3.8) we obtain $\dot{\sigma}(t)$ in the form

$$\dot{\sigma}(t) = (\dot{P}_m(t) - <\dot{P}_m(t)>_q^t)F_m(t) + (P_m(t) - <P_m(t)>_q^t)\frac{\partial F_m(t)}{\partial t} \tag{3.12}$$

where we take into account the identity

$$<\dot{P}_m(t)>_q^t F_m(t) = \mathrm{Sp}[i\mathcal{L}\rho_q(t)] = 0 \tag{3.13}$$

With the use of the conditions (2.4) the equations for $\frac{\partial F_m(t)}{\partial t}$ can be written as

$$\frac{\partial F_m(t)}{\partial t} = \frac{\delta F_m(t)}{\delta <P_n(t)>_q^t}\frac{\partial <P_n(t)>}{\partial t} \tag{3.14}$$

There is the function $\rho(t)$ on the right–hand–side of these equations. So equations (3.11) and (3.14) form a closed set of integrodifferential equations to determine the nonequilibrium distribution function of the contracted description, $\rho(t)$, and the time-dependent thermodynamic parameters $F_m(t)$ of this contracted description.

Thus, the basis of the nonequilibrium statistical operator method is formed by the set of the equations containing the dynamic equation (3.3) or (3.11) for $\rho(t)$ and the macroscopic evolution equation (3.14) for the thermodynamic parameters $F_m(t)$. The last are referred as hydrodynamic equations, because they form a familiar set of hydrodynamic equations in the particular case of a weakly nonequilibrium fluid, using equation (2.14) as $\rho_q(t)$.

Substituting equations (3.12) and (3.14) into equation (3.11) we get the integral equation for $\rho(t)$ in the form

$$\rho(t) = \rho_q(t) + \int_{\infty}^{t} dt_1 U_\varepsilon(t,t_1)\rho_q(t_1)\Big\{\left(\dot{P}_m(t_1) - < \dot{P}_m(t_1) >_q^t\right) F_m(t_1)$$

$$+(P_m(t_1) - < P_m(t_1) >_q^{t_1}) \left(\delta F_m(t_1)/\delta < P_n(t_1) >_q^{t_1}\right)$$

$$\times \mathrm{Sp}\left[\dot{P}_n(t_1)\rho(t_1)\right]\Big\} \tag{3.15}$$

There are no principal difficulties in constructing a solution to this equation in the form of the von Neumann series. However subsequent analysis of this series may be easily performed for each specific system, but not for a general case.

In particular, if the state of a system is not too far from the equilibrium state, and the function $\rho(t)$ differs slightly from $\rho_q(t)$ it will suffice to confine the solution results from the first iteration, with $\rho(t) = \rho_q(t)$, on the right–hand–part of equation (3.15).

In the subsequent chapters of this book the nonequilibrium distribution function found with this approach will be used to derive transport equations and analyze fluctuation processes in heterogeneous systems fluid.

Note that the field of applicability of the Zubarev method is significantly larger than the range of topics discussed in the present book. In addition to the original monograph by Zubarev [4] one should recommend the constructive review [6] of the recent investigations on the topic.

Chapter 2

Transport Phenomena at Interfaces and the Problem of Boundary Conditions in Hydrodynamics

2.1 Introduction

In phenomenological nonequilibrium thermodynamics, transport processes in bulk phases (gaseous, liquid, or solid) can be described (see, for example, [7]) by means of the well known equations of diffusion, heat conduction, the Navier-Stokes equations, etc. On the boundaries of the bulk phases it is customary to invoke simple physical arguments to specify the values of the corresponding quantities (a.g., concentration, temperature, velocity, etc.) or their gradients.

However, in the most general cases, when one is dealing with surface transport of mass, energy, or momentum along the interface, the boundary itself must be regarded as an independent phase, and the general methods of nonequilibrium thermodynamics are to be used for description. This was first noted by Waldmann [8] in a study of the boundary between two immiscible fluids, where singular (i.e., with nonvanishing surface density) flows of energy and momentum may occur. He obtained an expression for the surface entropy production, from which he derived the linear relationships of nonequilibrium thermodynamics between the flows along and across the surface and the values of the temperatures and velocities of the bulk phases on either sides of the boundary. In most general form, the nonequilibrium thermodynamics of such boundaries with surface mass, energy, and momentum densities were investigated by Bedeaux, Albano and Mazur [9, 10] and Kovac [11].

On the other hand, the methods of statistical mechanics developed so far provide, on the basis of first principles, all equations of the nonequilibrium thermodynamics of the bulk phases as well as explicit expressions for the transport coefficients in terms of molecular parameters. For rarefied systems, this is usually done by means of the Boltzmann equation, and for dense systems by the construction of an appropriate particular solution to the Liouville equation (see Zubarev monograph). Hitherto, transport processes on the interface have not been studied in statistical mechanics except for the boundary between a rarefied gas and a solid (this was done using the Boltzmann equation method).

In the present chapter, we consider nonequilibrium phenomena on the surface separating two immiscible fluids, including a gas–solid surface; a gas–liquid surface; a liquid–liquid surface, and a liquid–solid surface.

We begin by constructing the nonequilibrium statistical thermodynamics of the transport phenomena on interfaces, taking into account the possibility of formation of a surface film with nonvanishing surface densities of the mass, energy, and momentum. We derive conservation laws and equations of ideal and nonideal hydrodynamics for such a system. Then, the special case of an interphase surface is considered that does not include surface densities of the mass, energy, and momentum. Such surfaces do not have intrinsic dynamics; therefore, their hydrodynamic equations degenerate into the system of boundary conditions for the hydrodynamic equations of the bulk phases.

The boundary conditions deduced in this manner are compared with the ones proposed in the phenomenological approach of Goodrich [12] and are used for investigation of capillary waves damping on a viscous liquid surface with the film of a surface-active substance.

2.2 Geometric Description of the Surface

Unlike the macroscopic approach, in the statistical theory, one must allow for the finite thickness of the surface layer, which is of the order of several molecular layers. In accordance with such a picture, two-phase systems are subdivided into three regions: **I** and **II** — the homogeneous bulk phases, and

S — the surface phase, in which all the transient phenomena occur. The boundaries Σ_{I} and Σ_{II} between phases **I** and II and phase **S** are specified by means of two functions

$$f^{\mathrm{I}}(\mathbf{x},t) \begin{array}{lll} > 0 & \text{if} & \mathbf{x}\in \mathbf{I} \\ = 0 & \text{if} & \mathbf{x}\in\Sigma_{\mathrm{I}} \\ < 0 & \text{if} & \mathbf{x}\in\mathbf{S}\end{array} \qquad f^{\mathrm{II}}(\mathbf{x},t) \begin{array}{lll} > 0 & \text{if} & \mathbf{x}\in \mathrm{II} \\ = 0 & \text{if} & \mathbf{x}\in\Sigma_{\mathrm{II}} \\ < 0 & \text{if} & \mathbf{x}\in\mathbf{S}\end{array} \tag{2.1}$$

Vectors $\mathbf{n}^{\mathrm{I}}$ and $\mathbf{n}^{\mathrm{II}}$ to Σ_{I} and Σ_{II} are defined as

$$\mathbf{n}^{\mathrm{I,II}} = \left.\frac{\mathrm{grad} f^{\mathrm{I,II}}}{|\mathrm{grad} f^{\mathrm{I,II}}|}\right|_{\Sigma_{\mathrm{I,II}}} \tag{2.2}$$

Since the functions f_{I} and f_{II} are constant (and equal zero) at the surfaces Σ_{I} and Σ_{II}, respectively, their time derivatives have the form

$$\frac{df^{I,\mathrm{II}}}{dt} \equiv \frac{\partial f^{\mathrm{I,II}}}{\partial t} + \mathbf{v}_{\Sigma_{\mathrm{I,II}}} \nabla f^{\mathrm{I,II}} = \frac{\partial f^{\mathrm{I,II}}}{\partial t} + \mathbf{n}^{\mathrm{I,II}}\mathbf{v}_{\Sigma_{\mathrm{I,II}}} |\mathrm{grad} f^{\mathrm{I,II}}| = 0 \tag{2.3}$$

Functions f_{I} and f_{II} are used for introducing the step functions:

$$\Theta^{\mathrm{I}}[f^{\mathrm{I}}(\mathbf{x},t)] = \begin{cases} 1 & \text{if } f^{\mathrm{I}} > 0 \\ 0 & \text{if } f^{\mathrm{I}} \le 0 \end{cases} \qquad \Theta^{\mathrm{II}}[f^{\mathrm{II}}(\mathbf{x},t)] = \begin{cases} 1 & \text{if } f^{\mathrm{II}} > 0 \\ 0 & \text{if } f^{\mathrm{II}} \le 0 \end{cases} \tag{2.4}$$

$$\Theta^{S} = 1 - \Theta^{\mathrm{I}} - \Theta^{\mathrm{II}} \tag{2.5}$$

As a consequence of equations (2.3), the derivatives of the Θ functions are

$$\frac{\partial}{\partial \mathbf{x}}\Theta^{\mathrm{I,II}}[f^{I.\mathrm{II}}(\mathbf{x},t)] = \frac{\partial f^{\mathrm{I,II}}}{\partial \mathbf{x}}\delta_{+}[f^{\mathrm{I,II}}(\mathbf{x},t)] = \mathbf{n}^{\mathrm{I,II}}\delta_{\Sigma_{I.\mathrm{II}}} \tag{2.6}$$

$$\frac{\partial}{\partial t}\Theta^{\mathrm{I,II}}[f^{\mathrm{I,II}}(\mathbf{x},t)] = \frac{\partial f^{\mathrm{I,II}}}{\partial t}\delta_{+}[f^{\mathrm{I,II}}(\mathbf{x},t)] = -\mathbf{n}^{\mathrm{I,II}}\delta_{\Sigma_{I.\mathrm{II}}}\mathbf{v}_{\Sigma_{\mathrm{I,II}}} \tag{2.7}$$

where

$$\delta_{\Sigma_{I.\mathrm{II}}} = |gradf^{\mathrm{I,II}}|\ \delta_{+}[f^{\mathrm{I,II}}(\mathbf{x},t)] \tag{2.8}$$

are surfaces δ_{+} functions that are nonzero on the outer sides of surfaces Σ_{I} and Σ_{II}, i.e., those facing the bulk phases **I** and II, respectively. Henceforth, the thickness of the surface layer will be regarded as constant along the layer as well as in time, which allows setting $\mathbf{n}^I\mathbf{v}_{\Sigma_I} = \mathbf{n}^{II}\mathbf{v}_{\Sigma_{II}} = v_{nS}$. If, in addition, we confine ourselves to the case of thin layers with thickness λ much smaller than the radius of curvature, then $\mathbf{n}^{II} = -\mathbf{n}^{I} = \mathbf{n}$.

The function Θ^S will be used frequently for integrating, along the normal $\mathbf{n}$, quantities describing the surface layer. By the definition of this function, such integrals in the local coordinate system whose z axis points along $\mathbf{n}$ and whose origin lies inside the surface layer should be understood as the limits

$$\int_{-\infty}^{\infty} dz\varphi(\mathbf{x}, t)\Theta^S(\mathbf{x}, t) = \lim_{\epsilon\to 0}\int_{-\lambda/2-\epsilon}^{\lambda/2+\epsilon} dz\varphi(\mathbf{x}, t) \equiv \tilde{\varphi}(\mathbf{s}, t) \tag{2.9}$$

where $\varphi(\mathbf{x}, t)$ is an arbitrary integrable function of $\mathbf{x}$ and t, and $\mathbf{s}$ is a two–dimensional vector.

2.3 Local Equilibrium Distribution

The hydrodynamically nonequilibrium state of a system is described by a set of thermodynamic properties such as temperature $T(\mathbf{x}, t)$ or $\beta(\mathbf{x}, t) = 1/kT(\mathbf{x}, t)$, average velocity $\mathbf{v}(\mathbf{x}, t)$, and chemical potentials $\mu_\alpha(\mathbf{x}, t)$ of the various components. In a surface system these parameters may be discontinuous at the boundaries $\Sigma_{\rm I}$ and $\Sigma_{\rm II}$. So, it is convenient to describe each phase in terms of its own set of properties $\beta^l(\mathbf{x}, t) = 1/kT^l(\mathbf{x},\ t)$, $\mathbf{v}^l(\mathbf{x}, t)$, and $\nu^l_\alpha(\mathbf{x}, t) = \beta^l(\mathbf{x}, t)\mu^l_\alpha(\mathbf{x}, t)$, ($l$= I, II, S). Each of these properties (variables) is a continuous function of position $x \in l$ and time. Because the surface layer is infinitesimally thin from the macroscopic point of view, and is described by the surface densities of energy, momentum and mass, its thermodynamic parameters $\beta^S, \mathbf{v}^S$, and ν^S_α can be regarded as constant along the normal to the layer. It is possible, however, to have discontinuities of the thermodynamic parameters at the interfaces $\Sigma_{\rm I}$, $\Sigma_{\rm II}$; such discontinuities are well known, e.g., at the boundary of a gas and a solid where $(\mathbf{v}^{\rm I} - \mathbf{v}^{\rm II})|_\Sigma \neq 0$ in the case of a velocity slip and $(\beta^{\rm I} - \beta^{\rm II})|_\Sigma \neq 0$ in the case of a temperature slip.

To construct the statistical mechanics of such a system the microscopic description of the system is to be taken as a starting point.

The Hamiltonian of the multicomponent system under study is

$$\mathcal{H} = \sum_{\alpha i}\left(\frac{p^2_{\alpha i}}{2m_\alpha} + \frac{1}{2}\sum_{\beta j} u^{\alpha\beta}(\mathbf{r}_{\alpha i} - \mathbf{r}_{\beta j})\right) \tag{3.1}$$

where $\mathbf{p}_{\alpha i}$ and $\mathbf{r}_{\alpha i}$ are the momentum and the coordinate of the i-th molecule of α component; m_α is the mass of a molecule of the α component; and $u^{\alpha\beta}$ is the potential of the intermolecular interaction.

As parameters $\{P_m\}$ of a reduced description, we take the dynamic densities of energy,

$$H(\mathbf{x}) = \sum_{\alpha i} \left(\frac{p_{\alpha i}^2}{2m_\alpha} + \frac{1}{2} \sum_{\beta j} u^{\alpha\beta}(\mathbf{x} - \mathbf{r}_{\beta j}) \right) \delta(\mathbf{x} - \mathbf{r}_{\alpha i}) \tag{3.2}$$

momentum ,

$$\mathbf{p}(\mathbf{x}) = \sum_{\alpha i} \mathbf{p}_{\alpha i} \delta(\mathbf{x} - \mathbf{r}_{\alpha i}) \tag{3.3}$$

and number of particles of the α component

$$n_\alpha(\mathbf{x}) = \sum_i \delta(\mathbf{x} - \mathbf{r}_{\alpha i}) \tag{3.4}$$

Then the quasiequilibrium distribution function of such a three-phase system can be written in the form

$$\rho_q(t) = Q_q^{-1} \exp\Big\{ - \int d\mathbf{x} \sum_{l=\mathrm{I,II,S}} \Theta^l(\mathbf{x}, t) \Big[\beta^l(\mathbf{x}, t) H(\mathbf{x}) - \beta^l(\mathbf{x}, t) \mathbf{v}^l(\mathbf{x}, t) \mathbf{p}(\mathbf{x})$$
$$- \sum_\alpha \left(\nu_\alpha^l(\mathbf{x}, t) - \beta^l(\mathbf{x}, t) m_\alpha \frac{(v^l)^2(\mathbf{x}, t)}{2} \right) n_\alpha(\mathbf{x}) \Big] \Big\} \tag{3.5}$$

To determine the thermodynamic parameters $\beta^S, \mathbf{v}^S$, and ν_α^S, we use equations (1.2.4)

$$< H(\mathbf{x}) >_q = < H(\mathbf{x}) > \equiv u(\mathbf{x}, t) + \beta^l(\mathbf{x}, t) \frac{(v^l)^2(\mathbf{x}, t)}{2} \tag{3.6}$$

$$< \mathbf{p}(\mathbf{x}) >_q = < \mathbf{p}(\mathbf{x}) > \equiv \varrho(\mathbf{x}, t) v^l(\mathbf{x}) \tag{3.7}$$

$$< n_\alpha(\mathbf{x}) >_q = < n_\alpha(\mathbf{x}) > \equiv \varrho_\alpha(\mathbf{x}, t)/m_\alpha \tag{3.8}$$

separately for $\mathbf{x} \in \mathbf{I}$, $\mathbf{x} \in \mathbf{II}$, and $\mathbf{x} \in \mathbf{S}$ for all values of t. Hence $< ... >_q$ represents averaging over the entire phase space with distribution function $\rho_q(t)$, and $< ... >$ is the averaging with the exact distribution function $\rho(t)$, from which the observable thermodynamic quantities, i.e., the internal energy density $u(\mathbf{x}, t)$, density of the α component $\varrho_\alpha(\mathbf{x}, t)$ and the momentum density $\varrho(\mathbf{x}, t)\mathbf{v}^l(\mathbf{x}, t)$ can be calculated. Thus, parameters β^S, $\mathbf{v}^S$, and ν_α^S for each time t are so selected that evaluation of the averages with the distribution function $\rho_q(t)$ yields average densities of energy, momentum, and masses of the components identical to their exact values.

It should be noted that if Θ–functions were not introduced in the definition of the function $\rho_q(t)$, equation (3.5), the quasiequilibrium distribution becomes the usual local equilibrium function (1.2.14) but its parameters β, $\mathbf{v}$, and ν undergo discontinuities on the boundaries. The introduction of the step Θ–functions permits the use of continuous thermodynamic parameters $\beta^l, \mathbf{v}^l, \nu^l_\alpha$, rather than piecewise continuous functions.

2.4 Gibbs Adsorption Equation

Just as the Gibbs equilibrium distribution function ensures correct thermodynamic description of an equilibrium system, we have the quasiequilibrium distribution of equation (3.5).

Let us consider the quasiequilibrium partition function Q_q. Omitting for brevity the arguments of $\beta^l, \mathbf{v}^l, \nu^l_\alpha$, we represent it in the form

$$Q_q^{-1} = \mathrm{Sp}\Bigg[\exp\Big\{-\int\limits_{V^{\mathrm{I}}} d\mathbf{x}\Big[\beta^{\mathrm{I}} H(\mathbf{x}) - \beta^{\mathrm{I}}\mathbf{v}^{\mathrm{I}}\mathbf{p}(\mathbf{x})$$
$$-\sum_\alpha \left(\nu^{\mathrm{I}}_\alpha - \beta^{\mathrm{I}} m_\alpha \frac{(v^{\mathrm{I}})^2}{2}\right) n_\alpha(\mathbf{x})\Big]$$
$$-\int\limits_{V^{\mathrm{II}}} d\mathbf{x}\Big[\beta^{\mathrm{II}} H(\mathbf{x}) - \beta^{\mathrm{II}}\mathbf{v}^{\mathrm{II}}\mathbf{p}(\mathbf{x}) - \sum_\alpha \left(\nu^{\mathrm{II}}_\alpha - \beta^{\mathrm{II}} m_\alpha \frac{(v^{\mathrm{II}})^2}{2}\right) n_\alpha(\mathbf{x})\Big]$$
$$-\int\limits_{A} d\mathbf{s}\Big[\beta^S \tilde{H}(\mathbf{s}) - \beta^S\mathbf{v}^S\tilde{\mathbf{p}}(\mathbf{s}) - \sum_\alpha \left(\nu^l_\alpha - \beta^S m_\alpha \frac{(v^S)^2}{2}\right) \tilde{n}_\alpha(\mathbf{s})\Big]\Big\}\Bigg] \tag{4.1}$$

Rather than the Θ–functions, we have now the constraints on the domains of integration over the volumes $V^{\mathrm{I}}(t)$ and $V^{\mathrm{II}}(t)$ of phases **I** and II.The volume integral over the phase **S** is represented, after integration, across the layer in the form of the surface integral over the surface $A(t)$, which is situated between Σ_{I} and Σ_{II} (a more definite specification of its position is not required here). The *tilde* is used to indicate the surface densities of the type equation (2.9). Thus, hereafter the superscript S will be omitted if the value is depicted with the tilde over the variable.

Varying the partition function (4.1), we find

$$\frac{\delta \ln Q_q}{\delta \beta^{\mathrm{I,II}}} = -u^{\mathrm{I,II}} + \sum_\alpha \mu^{\mathrm{I,II}}_\alpha \varrho^{\mathrm{I,II}}_\alpha / m_\alpha, \quad \frac{\delta \ln Q_q}{\delta \beta^S} = -\tilde{u} + \sum_\alpha \mu^S_\alpha \tilde{\varrho}_\alpha / m_\alpha \tag{4.2}$$

$$\frac{\delta \ln Q_q}{\delta \mu^{\mathrm{I,II}}_\alpha} = \beta^{\mathrm{I,II}} \varrho^{\mathrm{I,II}}_\alpha / m_\alpha, \quad \frac{\delta \ln Q_q}{\delta \mu^S_\alpha} = \beta^S \tilde{\varrho}_\alpha / m_\alpha, \quad \frac{\delta \ln Q_q}{\delta \mathbf{v}^l} = 0 \tag{4.3}$$

From these relations, for the total variation of $\delta \ln Q_q$, we obtain

$$\delta \ln Q_q = \int\limits_{V^{\mathrm{I}}} d\mathbf{x} \left\{ - \left[u^{\mathrm{I}}(\mathbf{x}, t) - \sum_\alpha \mu_\alpha^{\mathrm{I}} \frac{\varrho_\alpha^{\mathrm{I}}(\mathbf{x}, t)}{m_\alpha} \right] \delta\beta^{\mathrm{I}} + \beta^{\mathrm{I}} \sum_\alpha \frac{\varrho_\alpha^{\mathrm{I}}(\mathbf{x}, t)}{m_\alpha} \delta\mu_\alpha^{\mathrm{I}} \right\}$$
$$+ \int\limits_{V^{\mathrm{II}}} d\mathbf{x} \left\{ - \left[u^{\mathrm{II}}(\mathbf{x}, t) - \sum_\alpha \mu_\alpha^{\mathrm{II}} \frac{\varrho_\alpha^{\mathrm{II}}(\mathbf{x}, t)}{m_\alpha} \right] \delta\beta^{\mathrm{II}} + \beta^{\mathrm{II}} \sum_\alpha \frac{\varrho_\alpha^{\mathrm{II}}(\mathbf{x}, t)}{m_\alpha} \delta\mu_\alpha^{\mathrm{II}} \right\}$$
$$+ \int\limits_{A} d\mathbf{s} \left\{ - \left[\tilde{u}(\mathbf{s}, t) - \sum_\alpha \mu_\alpha^{S} \frac{\tilde{\varrho}_\alpha(\mathbf{s}, t)}{m_\alpha} \right] \delta\beta^{\mathrm{I}} + \beta^{S} \sum_\alpha \frac{\tilde{\varrho}_\alpha^{\mathrm{I}}(\mathbf{x}, t)}{m_\alpha} \delta\mu_\alpha^{S} \right\} \tag{4.4}$$

On the other hand, $\ln Q_q$ can be represented in the form

$$\ln Q_q = -\int_{V^{\mathrm{I}}} d\mathbf{x}\beta^{\mathrm{I}}\Omega^{\mathrm{I}}(\mathbf{x}, t) - \int_{V^{\mathrm{II}}} d\mathbf{x}\beta^{\mathrm{II}}\Omega^{\mathrm{II}}(\mathbf{x}, t) - \int_A d\mathbf{s}\beta^S\Omega^S(\mathbf{s}, t) \tag{4.5}$$

where we have introduced the thermodynamic potentials $\Omega^{\mathrm{I,II}}, \Omega^S$, which can be defined formally in terms of the thermodynamic potential per particle in the neighborhood of the point $\mathbf{x}$ in the equilibrium ensemble characterized by parameters $\beta^l(\mathbf{x})$, and $\nu_\alpha^l(\mathbf{x})$, multiplied by the mean number density of the particles $\sum_\alpha \varrho^{\mathrm{I,II}}(\mathbf{x})/m_\alpha$ or $\sum_\alpha \tilde{\varrho}(\mathbf{s})/m_\alpha$. Since the only external parameters are the volumes V^{I}, and V^{II}, and the surface A (the radius of curvature here assumed to be considerably greater than the film thickness), it follows from the definition of the thermodynamic potential that

$$\Omega^{\mathrm{I,II}}(\mathbf{x}, t) = -p^{\mathrm{I,II}}(\mathbf{x}, t), \qquad \Omega^S(\mathbf{s}, t) = \gamma(\mathbf{s}, t) \tag{4.6}$$

where $p^{\mathrm{I}}(\mathbf{x}, t)$ and $p^{\mathrm{II}}(\mathbf{x}, t)$ are the local pressures in phases **I** and II, and $\gamma(\mathbf{s}, t)$ is the local surface tension of the interphase film.

On the basis of the entropy definition $S = -\mathrm{Sp}(\rho_q \ln \rho_q)$, we introduce the entropy density in the general form

$$S^l(\mathbf{x}, t) = F_m^l(\mathbf{x}, t) < P_m(\mathbf{x}) > -\Omega^l(\mathbf{x}, t) \tag{4.7}$$

and in particular forms for the bulk and surface phases of our system

$$S^{I,l} = \beta^{\mathrm{I,II}} \left[u^{\mathrm{I,II}}(\mathbf{x}, t) - \sum_\alpha \mu_\alpha^{\mathrm{I,II}} \frac{\varrho_\alpha^{\mathrm{I,II}}(\mathbf{x}, t)}{m_\alpha} - \Omega^{I,l}(\mathbf{x}, t) \right] \tag{4.8}$$

$$\tilde{S} = \beta^S \left[\tilde{u}(\mathbf{s}, t) - \sum_\alpha \mu_\alpha^S \frac{\tilde{\varrho}_\alpha(\mathbf{s}, t)}{m_\alpha} - \Omega^S(\mathbf{s}, t) \right] \tag{4.9}$$

From equation (4.7) we get the main thermodynamic relations for each phase

$$F^l_m = \frac{\partial S^l}{\partial < P_m >_q} \tag{4.10}$$

Comparing with equation (4.4) the variation of expression (4.5) and using the arbitrariness of the volumes V^0 and $V^{\rm II}$ and the surface A, we obtain the Gibbs–Duhem relation for the bulk phases

$$\delta(\beta^{\rm I,II}\Omega^{\rm I,II}(\mathbf{x},t)) = u^{\rm I,II}(\mathbf{x},t)\delta\beta^{\rm I,II} - \sum_\alpha \frac{\varrho^{\rm I,II}_\alpha(\mathbf{x},t)}{m_\alpha}\delta(\beta^{\rm I,II}\mu^{\rm I,II}_\alpha) \tag{4.11}$$

and a similar relation for the surface phase:

$$\delta(\beta^S\Omega^S(\mathbf{s},t)) = \tilde{u}(\mathbf{s},t)\delta\beta^S - \sum_\alpha \frac{\tilde{\varrho}_\alpha(\mathbf{s},t)}{m_\alpha}\delta(\beta^S\mu^S_\alpha) \tag{4.12}$$

Using equations (4.8) and (4.9) for the entropy densities, these relations may be rewritten in the form

$$-\delta p^{\rm I,II}(\mathbf{x},t) + S^{\rm I,II}(\mathbf{x},t)\delta(\beta^{\rm I,II})^{-1} + \sum_\alpha \frac{\varrho^{\rm I,II}_\alpha(\mathbf{x},t)}{m_\alpha}\delta\mu^{\rm I,II}_\alpha(\mathbf{x},t) = 0 \tag{4.13}$$

$$\delta\gamma(\mathbf{s},t)) + \tilde{S}(\mathbf{s},t)\delta(\beta^S(\mathbf{s},t))^{-1} + \sum_\alpha \frac{\tilde{\varrho}_\alpha(\mathbf{s},t)}{m_\alpha}\delta\mu^S_\alpha(\mathbf{s},t) = 0 \tag{4.14}$$

The last equation is a local quasiequilibrium generalization of the Gibbs adsorption equation, which is the basis of the thermodynamic theory of surface phenomena (see, as an example, [13]).

2.5 Conservation Laws

In the statistical approach, equations of motion for the mean energy, momentum, and mass densities are obtained by averaging equations of motion for the densities $\dot{n}_\alpha(\mathbf{x})$, $\dot{\mathbf{p}}(\mathbf{x})$, and $\dot{H}(\mathbf{x})$, which are defined as the Poisson brackets, with the total Hamiltonian of the system H:

$$\dot{n}_\alpha(\mathbf{x}) = \{\mathcal{H}, n_\alpha(\mathbf{x})\} \tag{5.1}$$

$$\dot{\mathbf{p}}(\mathbf{x}) = \{\mathcal{H}, \mathbf{p}(\mathbf{x})\} \tag{5.2}$$

$$\dot{H}(\mathbf{x}) = \{\mathcal{H}, H(\mathbf{x})\} \tag{5.3}$$

Calculating the Poisson brackets, we represent the first of these equations in the form of local conservation law:

$$\dot{n}_\alpha(\mathbf{x}) = -\mathrm{div}\mathbf{j}_\alpha(\mathbf{x}) \tag{5.4}$$

where

$$\mathbf{j}_\alpha(\mathbf{x}) = \sum \frac{\mathbf{p}_{\alpha i}}{m_\alpha}\delta(\mathbf{x}-\mathbf{r}_{\alpha i}) \tag{5.5}$$

is the flow of the number density of the particles of component α.

Let us now consider in more details equation (5.2) for the momentum density. As a result of a direct calculation of the Poisson brackets, we get

$$\dot{\mathbf{p}}(\mathbf{x}) = \sum_{\alpha i}\left[\frac{\mathbf{p}_{\alpha i}\mathbf{p}_{\alpha i}}{m_{\alpha i}}\frac{\partial\delta(\mathbf{x}-\mathbf{r}_{\alpha i})}{\partial\mathbf{r}_{\alpha i}} + \frac{1}{2}\sum_{\beta j}\frac{\partial u_{ij}^{\alpha\beta}}{\partial\mathbf{r}_{\alpha i}}(\delta(\mathbf{x}-\mathbf{r}_{\alpha i}) - \delta(\mathbf{x}-\mathbf{r}_{\beta j}))\right] \tag{5.6}$$

Now we shall try to represent this equation in the form of a conservation law.

There is no problem in representing the first member in the right– hand–side of equation (5.6) as a divergence of a vector. Thus, we analyze the second one.

As the potential $u_{ij}^{\alpha\beta}$ is assumed to be short–range it is possible to expand one of the δ-functions in the right–hand– side of equation (5.6) in the Taylor series. This procedure is quite legitimate [14] for δ–functions if they are defined on a space of continuous functions possessing continuous derivatives of all orders. Then, expanding the function $\delta(\mathbf{x}-\mathbf{r}_{\beta j})$ on the difference $\mathbf{r}_{ij}^{\alpha\beta} = \mathbf{r}_{\alpha i} - \mathbf{r}_{\beta j}$ we could get the second term of equation (5.6) in the linear approximation as the divergence of a vector:

$$\sum_{\alpha i}\sum_{\beta j}\frac{\partial u_{ij}^{\alpha\beta}}{\partial\mathbf{r}_{\alpha i}}(\delta(\mathbf{x}-\mathbf{r}_{\alpha i}) - \delta(\mathbf{x}-\mathbf{r}_{\beta j})) = -\frac{\partial}{\partial\mathbf{x}}\sum_{\alpha i}\sum_{\beta j}\frac{\partial u_{ij}^{\alpha\beta}}{\partial\mathbf{r}_{\alpha i}}\mathbf{r}_{ij}^{\alpha\beta}\delta(\mathbf{x}-\mathbf{r}_{\alpha i}) \tag{5.7}$$

In our specific case δ-functions are defined on a space of piecewise continuous functions undergoing discontinuities on the boundary surfaces Σ_0 and Σ_{II}. Then the previous Taylor expansion of δ-functions is only valid if the coordinates $\mathbf{r}_{\alpha i}$ and $\mathbf{r}_{\beta j}$ both belong to the same phase **0** or II or **S**; that is, the vector $\mathbf{r}_{ij}^{\alpha\beta}$ does not intersect the surfaces Σ_0 or Σ_{II}. So, the second summation in equation (5.6) is to be confined to the molecules βj that are in the same phase as the molecules αi (this constraint on the summation will be hereafter denoted by the prime symbol). In the general case of the molecules αi and βj from different phases, the second term in equation

(5.6) cannot be represented as a divergence of a vector and such interactions appear in a momentum conservation law as a source or an outer force:

$$\dot{\mathbf{p}}(\mathbf{x}) = -\mathrm{Div}\mathbf{T}'(\mathbf{x}) + \mathbf{F}(\mathbf{x}) \tag{5.8}$$

where

$$\mathbf{T}'(\mathbf{x}) = \sum_{\alpha i}\left[\frac{\mathbf{p}_{\alpha i}\mathbf{p}_{\alpha i}}{m_{\alpha i}} + \frac{1}{2}\sum\nolimits'_{\beta j}\frac{\partial u_{ij}^{\alpha\beta}}{\partial \mathbf{r}_{\alpha i}}\mathbf{r}_{ij}^{\alpha\beta}\right]\delta(\mathbf{x}-\mathbf{r}_{\alpha i}) \tag{5.9}$$

is the tensor of the momentum flow in the phase where the coordinate $\mathbf{r}_{\alpha i}$ is located, and

$$\mathbf{F}(\mathbf{x}) = -\sum_{\alpha i}\sum\nolimits''_{\beta j}\frac{\partial u_{ij}^{\alpha\beta}}{\partial \mathbf{r}_{\alpha i}}\mathbf{r}_{ij}^{\alpha\beta}\delta(\mathbf{x}-\mathbf{r}_{\alpha i}) \tag{5.10}$$

is the force of interaction between phases, the two primes indicating that interactions with molecules βj from other phases only are taken into account.

In real situations of macroscopically large volumes of bulk phases **0** and II, it is possible to replace the Θ-functions describing parameter discontinuities, when present, on the phase boundaries by macroscopically equivalent and quite regular functions. Then equation (5.6) may be represented as the conservation law, that is,

$$\dot{\mathbf{p}}(\mathbf{x}) = -\mathrm{Div}\mathbf{T}(\mathbf{x}) \tag{5.11}$$

where

$$\mathbf{T}(\mathbf{x}) = \sum_{\alpha i}\left[\frac{\mathbf{p}_{\alpha i}\mathbf{p}_{\alpha i}}{m_{\alpha i}} + \frac{1}{2}\sum_{\beta j}\frac{\partial u_{ij}^{\alpha\beta}}{\partial \mathbf{r}_{\alpha i}}\mathbf{r}_{ij}^{\alpha\beta}\right]\delta(\mathbf{x}-\mathbf{r}_{\alpha i}) \tag{5.12}$$

is the momentum flow tensor in the system.

Similarly we get the conservation law for the energy density in the form

$$\dot{H}(\mathbf{x}) = -\mathrm{div}\mathbf{j}_H(\mathbf{x}) \tag{5.13}$$

where

$$\mathbf{j}_H(\mathbf{x}) = \sum_{\alpha i}\frac{\mathbf{p}_{\alpha i}}{m_{\alpha i}}\left[\frac{p_{\alpha i}^2}{2m_{\alpha i}} + \frac{1}{2}\sum_{\beta j}u_{ij}^{\alpha\beta} + \frac{1}{2}\sum_{\beta j}\frac{\partial u_{ij}^{\alpha\beta}}{\partial \mathbf{r}_{\alpha i}}\mathbf{r}_{ij}^{\alpha\beta}\right]\delta(\mathbf{x}-\mathbf{r}_{\alpha i}) \tag{5.14}$$

is the energy flow vector.

Averaging the resulting conservation laws, equations (5.4), (5.11), and (5.13), with a nonequilibrium distribution function $\rho(t)$, satisfying the Liouville equation yields the following equations of motion for the average densities:

$$\frac{1}{m_\alpha}\frac{\partial}{\partial t}\varrho_\alpha(\mathbf{x},t) = -\mathrm{div} < \mathbf{j}_\alpha(\mathbf{x}) > \tag{5.15}$$

$$\frac{\partial}{\partial t}\varrho_\alpha(\mathbf{x},t)\mathbf{v}^l(\mathbf{x},t) = -\mathrm{Div} < \mathbf{T}(\mathbf{x}) >> \tag{5.16}$$

$$\frac{\partial}{\partial t}\left[u(\mathbf{x},t) + \frac{1}{2}\varrho_\alpha(\mathbf{x},t)(v^l)^2(\mathbf{x},t)\right] = -\mathrm{div} < \mathbf{j}_H(\mathbf{x}) > \tag{5.17}$$

where superscript l takes the values **I**, II, or **S**, depending on the region in which the coordinate $\mathbf{x}$ is situated.

The transport equations for the bulk phase have already been well defined. Hence our entire analysis will pertain to the surface layer **S**. This layer is described by two-dimensional densities of the type in equation (2.8) rather than three-dimensional ones. The transport equations should be derived by integration of equations (5.15) through (5.17) across the layer **S**. Multiplying equation (5.15) by Θ^S, and integrating along normal to the layer in conjunction with equations (2.5) and (2.6) we get

$$\begin{aligned}\frac{1}{m_\alpha}\frac{\partial}{\partial t}\tilde{\varrho}_\alpha(\mathbf{x},t) = {} & \frac{1}{m_\alpha}\mathbf{n}\cdot\mathbf{v}^S(\mathbf{x},t)\big(\varrho_\alpha^{\mathrm{II}}|_{\Sigma_{\mathrm{II}}} - \varrho_\alpha^{\mathrm{I}}|_{\Sigma_{\mathrm{I}}}\big) - \mathrm{div}_\tau < \tilde{\mathbf{j}}_\alpha(\mathbf{x}) > \\ & -\mathbf{n}\cdot\big(< \mathbf{j}_\alpha(\mathbf{x}) > |_{\Sigma_{\mathrm{II}}} - < \mathbf{j}_\alpha(\mathbf{x}) > |_{\Sigma_{\mathrm{I}}}\big)\end{aligned} \tag{5.18}$$

where div_τ is the tangential divergence (see Appendix A).

Introducing diffusion flow

$$J_\alpha^l =< \mathbf{j}_\alpha(\mathbf{x}) > -\frac{1}{m_\alpha}\varrho_\alpha^l \tag{5.19}$$

we derive from equation (5.15) the conservation law for the surface density of mass of the α-component:

$$\begin{aligned}\frac{\partial}{\partial t}\tilde{\varrho}_\alpha + \mathbf{v}^S\cdot\nabla_\tau\tilde{\varrho}_\alpha = {} & -\tilde{\varrho}_\alpha\mathrm{div}_\tau\mathbf{v}^S - m_\alpha\mathrm{div}_\tau\tilde{\mathbf{J}}_\alpha \\ & -\mathbf{n}\cdot\big(m_\alpha\mathbf{J}_\alpha^{\mathrm{II}}|_{\Sigma_{\mathrm{II}}} - m_\alpha\mathbf{J}_\alpha^{\mathrm{I}}|_{\Sigma_{\mathrm{I}}}\big) \\ & -\mathbf{n}\cdot\big[\varrho_\alpha^{\mathrm{II}}(\mathbf{v}^{\mathrm{II}} - \mathbf{v}^S)|_{\Sigma_{\mathrm{II}}} - \varrho_\alpha^{\mathrm{I}}(\mathbf{v}^{\mathrm{I}} - \mathbf{v}^S)|_{\Sigma_{\mathrm{I}}}\big)\big]\end{aligned} \tag{5.20}$$

where ∇_τ is the two-dimensional tangential gradient (see Appendix A). It follows from the definition, equation (5.19), that $\sum_\alpha m_\alpha \mathbf{J}_\alpha = 0$. Hence, for the total density $\tilde{\varrho}(\mathbf{x}, t) = \sum_\alpha \tilde{\varrho}_\alpha(\mathbf{x}, t)$, we obtain from equation (5.20)

$$\begin{aligned} \frac{\partial}{\partial t}\tilde{\varrho} + \mathbf{v}^S \cdot \nabla_\tau \tilde{\varrho}_\alpha &= -\tilde{\varrho}\,\mathrm{div}_\tau \mathbf{v}^S \\ &\quad -\mathbf{n} \cdot [\varrho^{\mathrm{II}}(\mathbf{v}^{\mathrm{II}} - \mathbf{v}^S)|_{\Sigma_{\mathrm{II}}} - \varrho^{\mathrm{I}}\mathbf{v}^{\mathrm{I}} - \mathbf{v}^S)|_{\Sigma_{\mathrm{I}}})] \end{aligned} \tag{5.21}$$

Similarly, introducing the tensor of mean momentum flow in a moving frame of reference

$$\mathbf{\Pi}^l(\mathbf{x}, t) =< \mathbf{T}(\mathbf{x}) > -\varrho^l(\mathbf{x}, t)\mathbf{v}^l(\mathbf{x}, t)\mathbf{v}^l(\mathbf{x}, t)$$

or the stress tensor in a moving frame of reference

$$\sigma^l(\mathbf{x}, t) = -\mathbf{\Pi}^l(\mathbf{x}, t) \tag{5.22}$$

and the energy flow

$$\mathbf{J}_H^l(\mathbf{x}, t) =< \mathbf{j}_H(\mathbf{x}, t) > - \left(u^l(\mathbf{x}, t) + \frac{1}{2}\varrho^l(\mathbf{x}, t) v^{l^2}(\mathbf{x}, t) - \sigma^l(\mathbf{x}, t) \right) \mathbf{v}^l(\mathbf{x}, t) \tag{5.23}$$

we obtain from equations (5.16), (5.17) and (5.20) the conservation laws for the surface momentum and energy densities:

$$\begin{aligned} \tilde{\varrho}\frac{\partial}{\partial t}\mathbf{v}^S + \tilde{\varrho}\mathbf{v}^S \cdot \nabla_\tau \mathbf{v}^S &= \mathrm{Div}_\tau \tilde{\sigma} + \mathbf{n} \cdot (\sigma^{\mathrm{II}}|_{\Sigma_{\mathrm{II}}} - \sigma^{\mathrm{I}}|_{\Sigma_{\mathrm{I}}}) \\ &\quad -\mathbf{n} \cdot [\varrho^{\mathrm{II}}(\mathbf{v}^{\mathrm{II}} - \mathbf{v}^S)(\mathbf{v}^{\mathrm{II}} - \mathbf{v}^S)|_{\Sigma_{\mathrm{II}}} - \varrho^{\mathrm{I}}(\mathbf{v}^{\mathrm{I}} - \mathbf{v}^S)(\mathbf{v}^{\mathrm{I}} - \mathbf{v}^S)|_{\Sigma_{\mathrm{I}}})] \end{aligned} \tag{5.24}$$

$$\begin{aligned} \frac{\partial}{\partial t}\tilde{u} + \mathbf{v}^S \cdot \nabla_\tau \tilde{u} = & -\tilde{u}\,\mathrm{div}_\tau \mathbf{v}^S - \mathrm{div}_\tau \tilde{J}_H^S - \\ & +\tilde{\sigma} : \nabla_\tau \mathbf{v}^S - \mathbf{n} \cdot (\mathbf{J}_H^{\mathrm{II}}|_{\Sigma_{\mathrm{II}}} - \mathbf{J}_H^{\mathrm{I}}|_{\Sigma_{\mathrm{I}}}) \\ & +\mathbf{n} \cdot [\varrho^{\mathrm{II}}(\mathbf{v}^{\mathrm{II}} - \mathbf{v}^S)(\mathbf{v}^{\mathrm{II}} - \mathbf{v}^S)|_{\Sigma_{\mathrm{II}}} \\ & -\varrho^{\mathrm{I}}(\mathbf{v}^{\mathrm{I}} - \mathbf{v}^S)(\mathbf{v}^{\mathrm{I}} - \mathbf{v}^S)|_{\Sigma_{\mathrm{I}}})] \cdot \mathbf{v}^S \\ & -\mathbf{n} \cdot [(u^{\mathrm{II}} - \sigma^{\mathrm{II}} + \frac{1}{2}\varrho^{\mathrm{II}}((v^{\mathrm{II}})^2 - (v^S)^2))(\mathbf{v}^{\mathrm{II}} - \mathbf{v}^S)|_{\Sigma_{\mathrm{II}}} \\ & -(u^{\mathrm{I}} - \sigma^{\mathrm{I}} + \frac{1}{2}\varrho^{\mathrm{I}}((v^{\mathrm{I}})^2 - (v^S)^2))(\mathbf{v}^{\mathrm{I}} - \mathbf{v}^S)|_{\Sigma_{\mathrm{I}}}] \end{aligned} \tag{5.25}$$

Conservation laws (5.15)-(5.17) for the bulk phases **I** and II can be reduced to the standard form using equations (5.19), (5.22), and (5.23):

$$\frac{\partial}{\partial t}\varrho_\alpha^{\mathrm{I,II}} + \mathbf{v}^{\mathrm{I,II}} \cdot \nabla \varrho_\alpha^{\mathrm{I,II}} = -\varrho_\alpha^{\mathrm{I,II}} \mathrm{div} \mathbf{v}^{\mathrm{I,II}} - m_\alpha \mathrm{div} \mathbf{J}_\alpha^{\mathrm{I,II}} \tag{5.26}$$

$$\frac{\partial}{\partial t}\varrho^{\mathrm{I,II}} + \mathbf{v}^{\mathrm{I,II}} \cdot \nabla \varrho^{\mathrm{I,II}} = -\varrho^{\mathrm{I,II}} \mathrm{div} \mathbf{v}^{\mathrm{I,II}} \tag{5.27}$$

$$\varrho^{\mathrm{I,II}} \frac{\partial}{\partial t}\mathbf{v}^{\mathrm{I,II}} + \varrho^{\mathrm{I,II}} \mathbf{v}^{\mathrm{I,II}} \cdot \nabla \mathbf{v}^{\mathrm{I,II}} = +\mathrm{Div}\sigma^{\mathrm{I,II}} \tag{5.28}$$

$$\frac{\partial}{\partial t}u^{\mathrm{I,II}} + \mathbf{v}^{\mathrm{I,II}} \cdot \nabla u^{\mathrm{I,II}} = -u^{\mathrm{I,II}} \mathrm{div} \mathbf{v}^{\mathrm{I,II}} + \sigma^{\mathrm{I,II}} : \nabla \mathbf{v}^{\mathrm{I,II}} - \mathrm{div} J_H^S \tag{5.29}$$

Together with equations (5.20), (5.21), (5.24), and (5.25) they form a set of equations that describes transport in the entire volume occupied by the two–phase (or, more correctly, three–phase) system.

2.6 Equations of Ideal Hydrodynamics

To achieve closure of the above set of equations, one must determine the average flows $\mathbf{J}_\alpha^l, \sigma^l$ and $\mathbf{J}_H^l$ from equations (5.19), (5.22), and (5.23). The simplest way of solving this problem consists in using the quasiequilibrium distribution function (3.5) for evaluating the average quantities $< ... >$. Since the quasiequilibrium distribution does not account for dissipative processes in the system, this approximation yields equations of ideal hydrodynamics, with no allowance for diffusion, viscosity, thermal conductivity, etc. Indeed, for the density flow of the α-component, after substitution of equation (3.6), we have

$$< \mathbf{j}_\alpha >_q = \frac{1}{m_\alpha} < \mathbf{p}_\alpha(\mathbf{x}) >_q = \frac{1}{m_\alpha}\varrho_\alpha^l \mathbf{v}^l \tag{6.1}$$

whence

$$\mathbf{J}_{\alpha_q} = < \mathbf{j}_\alpha >_q - \frac{1}{m_\alpha}\varrho_\alpha^l \mathbf{v}^l = 0 \tag{6.2}$$

Thus, the equation of ideal hydrodynamics for the density of α-th component becomes

$$\begin{aligned} \frac{\partial}{\partial t}\tilde{\varrho}_\alpha + \mathbf{v}^S \cdot \nabla_\tau \tilde{\varrho}_\alpha &= -\tilde{\varrho}_\alpha^S \mathrm{div}_\tau \mathbf{v}^S \\ &-\mathbf{n} \cdot [\varrho_\alpha^{\mathrm{II}}(\mathbf{v}^{\mathrm{II}} - \mathbf{v}^S)|_{\Sigma_{\mathrm{II}}} - \varrho_\alpha^{\mathrm{I}}(\mathbf{v}^{\mathrm{I}} - \mathbf{v}^S)|_{\Sigma_{\mathrm{I}}})] \end{aligned} \tag{6.3}$$

and for full density

$$\frac{\partial}{\partial t}\tilde{\varrho} + \mathbf{v}^S \cdot \nabla_\tau \tilde{\varrho} = -\tilde{\varrho}\,\mathrm{div}_\tau \mathbf{v}^S$$
$$-\mathbf{n} \cdot [\varrho^{\mathrm{II}}(\mathbf{v}^{\mathrm{II}} - \mathbf{v}^S)|_{\Sigma_{\mathrm{II}}} - \varrho^{\mathrm{I}}(\mathbf{v}^{\mathrm{I}} - \mathbf{v}^S)|_{\Sigma_{\mathrm{I}}})] \tag{6.4}$$

Consider now the conservation equation for the momentum density. If $\mathbf{x} \in \mathbf{S}$, then averaging $< \ldots >_q$ in equation (5.22) leads, by virtue of the anisotropy of the layer, to different values of the pressure along and across the layer [15], as a result of which the pressure tensor in the local coordinate system has the form

$$\sigma_q^S(\mathbf{x}, t) = -p_n(\mathbf{x}, t)\mathbf{U} + (p_n(\mathbf{x}, t) - p_\tau(\mathbf{x}, t))\mathbf{U}_\tau \tag{6.5}$$

where $\mathbf{U}$ is a unit tensor, and $\mathbf{U}_\tau = \mathbf{U} - \mathbf{nn}$.

The pressure tensor of the surface phase results from the integration of equation (6.5) across the layer in the form

$$\tilde{\sigma}_q^S(\mathbf{x}, t) = -\tilde{p}_n(\mathbf{x}, t)\mathbf{U} + \gamma \mathbf{U}_\tau \tag{6.6}$$

Here

$$\gamma = \tilde{p}_n - \tilde{p}_\tau \equiv -\int_{\Sigma_{\mathrm{I}}}^{\Sigma_{\mathrm{II}}} dz(p_n - p_\tau) \tag{6.7}$$

is the surface tension coefficient [15] (see Appendix B). We introduce the surface tension tensor $\Gamma = \gamma \mathbf{U}_\tau$. Then, the momentum transport equation for the surface layer in the approximation of ideal hydrodynamics takes the form

$$\tilde{\varrho}\frac{\partial}{\partial t}\mathbf{v}^S + \tilde{\varrho}\mathbf{v}^S \cdot \nabla_\tau \mathbf{v}^S = -\nabla_\tau \tilde{-}\mathrm{Div}_\tau \Gamma - \mathbf{n}(p^{\mathrm{II}}|_{\Sigma_{\mathrm{II}}} - p^{\mathrm{I}}|_{\Sigma_{\mathrm{I}}})$$
$$-\mathbf{n} \cdot [\varrho^{\mathrm{II}}(\mathbf{v}^{\mathrm{II}} - \mathbf{v}^S)(\mathbf{v}^{\mathrm{II}} - \mathbf{v}^S)|_{\Sigma_{\mathrm{II}}} - \varrho^{\mathrm{I}}(\mathbf{v}^{\mathrm{I}} - \mathbf{v}^S)(\mathbf{v}^{\mathrm{I}} - \mathbf{v}^S)|_{\Sigma_{\mathrm{I}}})] \tag{6.8}$$

$$\begin{aligned}\frac{\partial}{\partial t}\tilde{u} + \mathbf{v}^S \cdot \nabla_\tau \tilde{u} = {} & -(\tilde{u} + \tilde{p}_n)\mathrm{div}_\tau \mathbf{v}^S \\ & +\Gamma : \nabla_\tau \mathbf{v}^S + \mathbf{n} \cdot [\varrho^{\mathrm{II}}(\mathbf{v}^{\mathrm{II}} - \mathbf{v}^S)(\mathbf{v}^{\mathrm{II}} - \mathbf{v}^S)|_{\Sigma_{\mathrm{II}}} \\ & -\varrho^{\mathrm{I}}(\mathbf{v}^{\mathrm{I}} - \mathbf{v}^S)(\mathbf{v}^{\mathrm{I}} - \mathbf{v}^S)|_{\Sigma_{\mathrm{I}}})] \cdot \mathbf{v}^S \\ & -\mathbf{n} \cdot [(h^{\mathrm{II}} + \frac{1}{2}\varrho^{\mathrm{II}}(v^{\mathrm{II}^2} - v^{S^2}))(\mathbf{v}^{\mathrm{II}} - \mathbf{v}^S)|_{\Sigma_{\mathrm{II}}} \\ & -(h^{\mathrm{I}} + \frac{1}{2}\varrho^{\mathrm{I}}(v^{I^2} - v^{S^2}))(\mathbf{v}^{\mathrm{I}} - \mathbf{v}^S)|_{\Sigma_{\mathrm{I}}}]\end{aligned} \tag{6.9}$$

where we have introduced the enthalpy densities of the bulk phases $h^{\mathrm{I,II}} = u^{\mathrm{I,II}} + p^{\mathrm{I,II}}$.

Related equations of ideal hydrodynamics for the bulk phases follow from equations (5.26) through (5.29) putting there $\mathbf{J}_\alpha^{\mathrm{I,II}} = 0$, $\sigma^{\mathrm{I,II}} = -p^{\mathrm{I,II}}\mathbf{U}$, $\mathbf{J}_H^{\mathrm{I,II}} = 0$. They are

$$\frac{\partial}{\partial t}\varrho_\alpha^{\mathrm{I,II}} + \mathbf{v}^{\mathrm{I,II}} \cdot \nabla\varrho_\alpha^{\mathrm{I,II}} = -\varrho_\alpha^{\mathrm{I,II}}\mathrm{div}\mathbf{v}^{\mathrm{I,II}} \tag{6.10}$$

$$\frac{\partial}{\partial t}\varrho^{\mathrm{I,II}} + \mathbf{v}^{\mathrm{I,II}} \cdot \nabla\varrho^{\mathrm{I,II}} = -\varrho^{\mathrm{I,II}}\mathrm{div}\mathbf{v}^{\mathrm{I,II}} \tag{6.11}$$

$$\varrho^{\mathrm{I,II}}\frac{\partial}{\partial t}\mathbf{v}^{\mathrm{I,II}} + \varrho^{\mathrm{I,II}}\mathbf{v}^{\mathrm{I,II}} \cdot \nabla\mathbf{v}^{\mathrm{I,II}} = -\nabla p^{\mathrm{I,II}} \tag{6.12}$$

$$\frac{\partial}{\partial t}u^{\mathrm{I,II}} + \mathbf{v}^{\mathrm{I,II}} \cdot \nabla u^{\mathrm{I,II}} = -h^{\mathrm{I,II}}\mathrm{div}\mathbf{v}^{\mathrm{I,II}} \tag{6.13}$$

The system of equations of ideal hydrodynamics *per se* is not closed. It is necessary then to add local state equations for the bulk phases

$$p^{\mathrm{I,II}} = p^{\mathrm{I,II}}(\varrho_\alpha^{\mathrm{I,II}}, u^{\mathrm{I,II}}, \mathbf{v}^{\mathrm{I,II}}) \tag{6.14}$$

and a related equation for the surface tension

$$\gamma = \gamma(\varrho_\alpha^S, u^S, \mathbf{v}^S) \tag{6.15}$$

The resultant set of equations describes the behavior of the complete three-phase system in the approximation of ideal hydrodynamics. To account for the dissipative effects associated with viscosity, diffusion, heat conduction, etc., it is necessary to calculate the mean flows $\mathbf{J}_\alpha^l, \sigma^l, \mathbf{J}_H^l$ with a more realistic distribution function than ρ_q.

2.7 Nonequilibrium Distribution Function

In accordance with the general fashion of the Zubarev method discussed in Chapter 1, we shall construct the nonequilibrium distribution function $\rho(t)$ as the solution to the Cauchy problem of the Liouville equation for the system with the Hamiltonian (3.1). It has the form

$$\begin{aligned}\rho(t) &= \varepsilon\int\limits_{-\infty}^{t} dt' e^{\varepsilon(t'-t)} e^{i\mathcal{L}(t'-t)}\rho_q(t')\\ &= \rho_q(t) - \int\limits_{-\infty}^{0} dt_1 e^{\varepsilon t_1} e^{i\mathcal{L}t_1}\left(\frac{\partial}{\partial t_1} + i\mathcal{L}\right)\rho_q(t+t_1)\end{aligned} \tag{7.1}$$

Further, it is necessary to calculate the integrand on the right-hand side of equation (7.1). We begin with the second term:

$$
\begin{aligned}
i\mathcal{L}\rho_q = -\rho_q \int d\mathbf{x} \sum_l \Theta^l \Big[& \beta^l(\mathbf{x},t) \left(\dot{H}(\mathbf{x}) - < \dot{H}(\mathbf{x}) >_q \right) \\
& - \beta^l(\mathbf{x},t) \mathbf{v}^l(\mathbf{x},t) \cdot \left(\dot{\mathbf{p}}(\mathbf{x}) - < \dot{\mathbf{p}}(\mathbf{x}) >_q \right) \\
& - \sum_\alpha (\nu_\alpha^l - \beta^l m_\alpha v^{l^2}/2) \left(\dot{n}_\alpha(\mathbf{x}) - < \dot{n}_\alpha(\mathbf{x}) >_q \right) \Big]
\end{aligned}
\tag{7.2}
$$

Using equations (5.4), (5.11), and (5.13), and integrating by parts we get equation (7.2) in the form

$$
\begin{aligned}
i\mathcal{L}\rho_q = -\rho_q \int d\mathbf{x} \sum_l \Theta^l \Big[& \nabla\beta^l \cdot \Big(\mathbf{j}_H - \mathbf{v}^l \cdot \mathbf{T}(\mathbf{x}) + \mathbf{p}(v^l)^2/2 \\
& - < \mathbf{j}_H - \mathbf{v}^l \cdot \mathbf{T}(\mathbf{x}) + \mathbf{p}(v^l)^2/2 >_q \Big) \\
& - \beta^l \nabla \mathbf{v}^l : \left(\mathbf{T}(\mathbf{x}) - \mathbf{p}\mathbf{v}^l - < \mathbf{T}(\mathbf{x}) - \mathbf{p}\mathbf{v}^l >_q \right) \\
& - \sum_\alpha \nabla \nu_\alpha^l \cdot \left(\mathbf{j}_\alpha(\mathbf{x}) - < \mathbf{j}_\alpha(\mathbf{x}) >_q \right) \Big] \\
& + \rho_q \int_{\Sigma_{\mathrm{I}}} d\mathbf{s}\mathbf{n} \cdot \Big[(\beta^{\mathrm{I}} - \beta^S) \Big(\mathbf{j}_H - \mathbf{v}^S \cdot \mathbf{T}(\mathbf{s}) + \mathbf{p}(v^S)^2/2 \\
& - < \mathbf{j}_H - \mathbf{v}^S \cdot \mathbf{T}(\mathbf{s}) + \mathbf{p}(v^S)^2/2 >_q \Big) \\
& - \beta^{\mathrm{I}} (\mathbf{v}^{\mathrm{I}} - \mathbf{v}^S) \cdot \left(\mathbf{T}(\mathbf{s}) - \mathbf{p}\mathbf{v}^S - < \mathbf{T}(\mathbf{s}) - \mathbf{p}\mathbf{v}^S >_q \right) \\
& - \sum_\alpha (\nu_\alpha^{\mathrm{I}} - \nu_\alpha^S + \beta^{\mathrm{I}} m_\alpha \frac{(\mathbf{v}^{\mathrm{I}} - \mathbf{v}^S)^2}{2}) \left(\mathbf{j}_\alpha(\mathbf{s}) - < \mathbf{j}_\alpha(\mathbf{s}) >_q \right) \Big] \\
& - \rho_q \int_{\Sigma_{\mathrm{II}}} d\mathbf{s}\mathbf{n} \cdot \Big[(\beta^{\mathrm{II}} - \beta^S) \Big(\mathbf{j}_H - \mathbf{v}^S \cdot \mathbf{T}(\mathbf{s}) + \mathbf{p}(v^S)^2/2 \\
& - < \mathbf{j}_H - \mathbf{v}^S \cdot \mathbf{T}(\mathbf{s}) + \mathbf{p}(v^S)^2/2 >_q \Big) \\
& - \beta^{\mathrm{II}} (\mathbf{v}^{\mathrm{II}} - \mathbf{v}^S) \cdot \left(\mathbf{T}(\mathbf{s}) - \mathbf{p}\mathbf{v}^S - < \mathbf{T}(\mathbf{s}) - \mathbf{p}\mathbf{v}^S >_q \right) \\
& - \sum_\alpha (\nu_\alpha^{\mathrm{II}} - \nu_\alpha^S + \beta^{\mathrm{II}} m_\alpha \frac{(\mathbf{v}^{\mathrm{II}} - \mathbf{v}^S)^2}{2}) \left(\mathbf{j}_\alpha(\mathbf{s}) - < \mathbf{j}_\alpha(\mathbf{s}) >_q \right) \Big]
\end{aligned}
\tag{7.3}
$$

The first term of the integrand of equation (7.1) may be rewritten using properties of the Θ-functions, equation (2.7), as

$$\begin{aligned}
\partial\rho_q/\partial t = &-\rho_q \int d\mathbf{x} \sum_l \Theta^l \Big[(\partial\beta^l/\partial t)\left(H_l'(\mathbf{x}) - u^l\right) - \beta^l (\partial\mathbf{v}^l/\partial t) \cdot \left(\mathbf{p}'(\mathbf{x}) - \varrho^l \mathbf{v}^l\right) \\
&- \sum_\alpha (\partial\nu_\alpha^l/\partial t)\left(n_\alpha(\mathbf{x}) - \varrho_\alpha^l/m_\alpha\right)\Big] \\
&-\rho_q \int_{\Sigma_\mathrm{I}} ds\ \mathbf{n}\cdot\mathbf{v}^S \Big[(\beta^\mathrm{I} - \beta^S)\Big(H_\mathrm{I}'(\mathbf{x}) - u^\mathrm{I} + (\mathbf{v}^\mathrm{I} - \mathbf{v}^S)\cdot\mathbf{p}_\mathrm{I}'(\mathbf{x}) \\
&- \sum_\alpha m_\alpha \frac{(v^\mathrm{I})^2 - (v^S)^2}{2}\left(n_\alpha(\mathbf{x}) - \varrho_\alpha^\mathrm{I}/m_\alpha\right)\Big) - \beta^\mathrm{I}(\mathbf{v}^\mathrm{I} - \mathbf{v}^S)\cdot\mathbf{p}_\mathrm{I}'(\mathbf{x}) \\
&- \sum_\alpha (\nu_\alpha^\mathrm{I} - \nu_\alpha^S + \beta^\mathrm{I} m_\alpha \frac{(\mathbf{v}^\mathrm{I} - \mathbf{v}^S)^2}{2})\left(n_\alpha(\mathbf{x}) - \varrho_\alpha^\mathrm{I}/m_\alpha\right)\Big] \\
&+\rho_q \int_{\Sigma_\mathrm{II}} ds\ \mathbf{n}\cdot\mathbf{v}^S \Big[(\beta^\mathrm{II} - \beta^S)\Big(H_\mathrm{II}'(\mathbf{x}) - u^\mathrm{II} + (\mathbf{v}^\mathrm{II} - \mathbf{v}^S)\cdot\mathbf{p}_\mathrm{II}'(\mathbf{x}) \\
&- \sum_\alpha m_\alpha \frac{(v^\mathrm{II})^2 - (v^S)^2}{2}\left(n_\alpha(\mathbf{x}) - \varrho_\alpha^\mathrm{II}/m_\alpha\right)\Big) - \beta^\mathrm{II}(\mathbf{v}^\mathrm{II} - \mathbf{v}^S)\cdot\mathbf{p}_\mathrm{II}'(\mathbf{x}) \\
&- \sum_\alpha (\nu_\alpha^\mathrm{I} - \nu_\alpha^S + \beta^\mathrm{II} m_\alpha \frac{(\mathbf{v}^\mathrm{II} - \mathbf{v}^S)^2}{2})\left(n_\alpha(\mathbf{x}) - \varrho_\alpha^\mathrm{II}/m_\alpha\right)\Big]
\end{aligned} \tag{7.4}$$

Here

$$H_l'(\mathbf{x}) = H(\mathbf{x}) - \mathbf{v}^l \cdot \mathbf{p}(\mathbf{x}) + \frac{v^{l^2}}{2} \sum_\alpha m_\alpha n_\alpha(\mathbf{x}) \tag{7.5}$$

$$\mathbf{p}_l'(\mathbf{x}) = \mathbf{p}(\mathbf{x}) - \mathbf{v}^l \sum_\alpha m_\alpha n_\alpha(\mathbf{x}) \tag{7.6}$$

are densities of energy and momentum in the local frame of reference moving with the velocity $\mathbf{v}^l$.

There are time derivatives of the thermodynamic parameters in equation (7.4), which can be eliminated. Here, as in the Chapman–Enskog approach when developing nonequilibrium thermodynamics, in the first approximation in parameters gradients the time derivatives are expressed in terms of space derivatives using the zeroth approximation.

In the framework of the developing approach it is suggested that the averages on the right–hand–side of equation (1.3.14) be calculated using the quasiequilibrium distribution function (3.5). This procedure is too cumbersome, so it is presented in detail in Appendix C. The main results of this procedure are

$$\frac{d\beta^{\rm I,II}}{dt} = \beta^{\rm I,II}\Big(\frac{\partial p^{\rm I,II}}{\partial u^{\rm I,II}}\Big)_{\varrho} {\rm div}\mathbf{v}^{\rm I,II} - \mathbf{v}^{\rm I,II}\cdot\nabla\beta^{\rm I,II} \tag{7.7}$$

$$\frac{d\mathbf{v}^{\rm I,II}}{dt} = \frac{u^{\rm I,II}+p^{\rm I,II}}{\beta^{\rm I,II}\varrho^{\rm I,II}}\nabla\beta^{\rm I,II} - \sum_\alpha \frac{\varrho_\alpha^{\rm I,II}}{m_\alpha\beta^{\rm I,II}\varrho^{\rm I,II}}\nabla\nu_\alpha^{\rm I,II} - \mathbf{v}^{\rm I,II}\cdot\nabla\mathbf{v}^{\rm I,II} \tag{7.8}$$

$$\frac{d\nu_\alpha^{\rm I,II}}{dt} = -\beta^{\rm I,II}\Big(\frac{\partial p^{\rm I,II}}{\partial \varrho_\alpha^{\rm I,II}}\Big)_u m_\alpha {\rm div}\mathbf{v}^{\rm I,II} - \mathbf{v}^{\rm I,II}\cdot\nabla\nu_\alpha^{\rm I,II} \tag{7.9}$$

$$\begin{aligned}\frac{d\beta^S}{dt} = {}& \beta^S\Big(\frac{\partial \tilde p_\tau^S}{\partial \tilde u}\Big)_{\tilde\varrho_\alpha} {\rm div}_\tau\mathbf{v}_\tau^S - \mathbf{v}^S\cdot\nabla\beta^S \\ & -\mathbf{n}\cdot\Big[\frac{\partial\beta^S}{\partial\tilde u}(h^{\rm II}\Delta\mathbf{v}^{\rm II} - h^{\rm I}\Delta\mathbf{v}^{\rm I}) + \sum_\alpha\frac{\partial\beta^S}{\partial\tilde\varrho_\alpha}(\varrho_\alpha^{\rm II}\Delta\mathbf{v}^{\rm II} - \varrho_\alpha^{\rm I}\Delta\mathbf{v}^{\rm I})\Big]\end{aligned} \tag{7.10}$$

$$\frac{d\mathbf{v}^S}{dt} = \frac{\tilde u + \tilde p_\tau}{\beta^S\tilde\varrho}\nabla_\tau beta^S - \sum_\alpha\frac{\tilde\varrho_\alpha}{m_\alpha\beta^S\tilde\varrho}\nabla_\tau\nu_\alpha^S - \mathbf{v}^S\cdot\nabla\mathbf{v}^S \tag{7.11}$$

$$\begin{aligned}\frac{d\nu_\alpha^S}{dt} = {}& -\beta^S\Big(\frac{\partial\tilde p_\tau}{\partial\tilde\varrho_\alpha}\Big)_{\tilde u} m_\alpha {\rm div}_\tau\mathbf{v}_\tau^S - \mathbf{v}^S\cdot\nabla\nu_\alpha^S \\ & -\mathbf{n}\cdot\Big[\frac{\partial\beta^S}{\partial\tilde\varrho_\alpha}(h^{\rm II}\Delta\mathbf{v}^{\rm II} - h^{\rm I}\Delta\mathbf{v}^{\rm I}) + \sum_{\alpha'}\frac{\partial\nu_{\alpha'}^S}{\partial\tilde\varrho_{\alpha'}}(\varrho_{\alpha'}^{\rm II}\Delta\mathbf{v}^{\rm II} - \varrho_{\alpha'}^{\rm I}\Delta\mathbf{v}^{\rm I})\Big]\end{aligned} \tag{7.12}$$

Thus, in linear approximation in the gradients of the thermodynamic parameters, and their discontinuities across the boundary surfaces, we obtain

$$\rho(t) = \rho_q(t)(1 + \Delta^1 + \Delta^2) \tag{7.13}$$

where

$$\begin{aligned}\Delta^1 = {}& \int d\mathbf{x}\int_{-\infty}^0 dt_1 e^{\varepsilon t_1}\sum_l \Theta^l(\mathbf{x}, t+t_1)\Big[\nabla\beta^l(\mathbf{x}, t+t_1)\cdot(\mathbf{j}_Q^l(\mathbf{x}, t_1) - <\mathbf{j}_Q^l>_q) \\ & -\beta^l(\mathbf{x}, t+t_1)\nabla\mathbf{v}^l(\mathbf{x}, t+t_1) : (\mathbf{j}_p^l(\mathbf{x}, t_1) - <\mathbf{j}_p^l>_q) \\ & -\sum_\alpha \nabla\nu_\alpha^l(\mathbf{x}, t+t_1)\cdot\mathbf{j}_{d\alpha}^l(\mathbf{x}, t_1)\Big]\end{aligned} \tag{7.14}$$

is the part of the nonequilibrium distribution function describing flows within each phase, and

$$
\begin{aligned}
\Delta^2 = &- \int\limits_{\Sigma_{\mathrm{I}}} d\mathbf{s} \int\limits_{\infty}^{0} dt_1 e^{\varepsilon t_1} \Big\{ \Delta\beta^{\mathrm{I}}(\mathbf{s}, t+t_1)\mathbf{n} \cdot (\mathbf{j}_Q^{\Sigma\mathrm{I}}(\mathbf{x}, t_1) - < \mathbf{j}_{\Sigma\mathrm{I}}^{l} >_q) \\
&- \beta^{\mathrm{I}}(\mathbf{x}, t+t_1)\Delta\mathbf{v}^{\mathrm{I}}(\mathbf{s}, t+t_1) : (\mathbf{j}_p^{\Sigma\mathrm{I}}(\mathbf{x}, t_1) - < \mathbf{j}_p^{\Sigma\mathrm{I}} >_q) \\
&- \sum_\alpha \Delta\nu_\alpha^{\mathrm{I}}(\mathbf{s}, t+t_1)\mathbf{n} \cdot \mathbf{j}_{d\alpha}^{\Sigma\mathrm{I}}(\mathbf{x}, t_1) \Big\} \\
&+ \int\limits_{\Sigma_1} d\mathbf{s} \int\limits_{-\infty}^{0} dt_1 e^{\varepsilon t_1} \Big\{ \Delta\beta^{\mathrm{II}}(\mathbf{s}, t+t_1)\mathbf{n} \cdot (\mathbf{j}_Q^{\Sigma\mathrm{II}}(\mathbf{x}, t_1) - < \mathbf{j}_{\Sigma\mathrm{II}}^{l} >_q) \\
&- \beta^{\mathrm{II}}(\mathbf{x}, t+t_1)\Delta\mathbf{v}^{\mathrm{II}}(\mathbf{s}, t+t_1) : (\mathbf{j}_p^{\Sigma\mathrm{II}}(\mathbf{x}, t_1) - < \mathbf{j}_p^{\Sigma\mathrm{II}} >_q) \\
&- \sum_\alpha \Delta\nu_\alpha^{\mathrm{II}}(\mathbf{s}, t+t_1)\mathbf{n} \cdot \mathbf{j}_{d\alpha}^{\Sigma\mathrm{II}}(\mathbf{x}, t_1) \Big\}
\end{aligned}
\tag{7.15}
$$

is the part of the distribution function that takes into account the transfer of energy, momentum, and particles between the surface phase and the bulk phases. In equation (7.14) we have used the usual [4] notions for heat, momentum, and diffusion flows in the phases $\mathbf{I}, \mathbf{II}, \mathbf{S}$, respectively, in the form

$$\mathbf{j}_Q(\mathbf{x}) = \mathbf{j}'_H(\mathbf{x}) - \frac{u+p}{\varrho}\mathbf{p}'(\mathbf{x}) \tag{7.16}$$

$$\mathbf{j}_p(\mathbf{x}) = T'(\mathbf{x}) - \left(\frac{\partial p}{\partial u}\right)_n H'(\mathbf{x})\mathbf{U} - \sum_\alpha \left(\frac{\partial p}{\partial \varrho_\alpha}\right)_u \varrho_\alpha(\mathbf{x})\mathbf{U} \tag{7.17}$$

$$\mathbf{j}_{d\alpha}(\mathbf{x}) = \mathbf{j}'_\alpha(\mathbf{x}) - \frac{\varrho_\alpha}{m_\alpha \varrho}\mathbf{p}'(\mathbf{x}) \tag{7.18}$$

Here for brevity the indexes **I** and II are omitted and equation (7.5) is used. The flows in the moving frame of reference are written as

$$T'(\mathbf{x}) = \mathbf{T}(\mathbf{x}) - \mathbf{v}(\mathbf{x})\mathbf{p}'(\mathbf{x}) - \mathbf{p}'(\mathbf{x})\mathbf{v}(\mathbf{x}) - \mathbf{v}(\mathbf{x})\mathbf{v}(\mathbf{x}) \sum_\alpha m_\alpha n_\alpha(\mathbf{x})\mathbf{v}(\mathbf{x}) \tag{7.19}$$

$$
\begin{aligned}
\mathbf{j}'_H(\mathbf{x}) = \mathbf{j}_H(\mathbf{x}) &- \big[H'(\mathbf{x}) + \frac{1}{2}v^2(\mathbf{x}) \sum_\alpha m_\alpha n_\alpha(\mathbf{x}) \\
&+ \mathbf{v}(\mathbf{x})\mathbf{p}'(\mathbf{x})\big]\mathbf{v}(\mathbf{x}) - \frac{1}{2}v^2(\mathbf{x})\mathbf{p}'(\mathbf{x}) - T'(\mathbf{x})\mathbf{v}(\mathbf{x})
\end{aligned}
\tag{7.20}
$$

$$\mathbf{j}'_\alpha(\mathbf{x}) = \mathbf{j}_\alpha(\mathbf{x}) - n_\alpha(\mathbf{x})\mathbf{v}(\mathbf{x}) \tag{7.21}$$

Similar expressions can be written for flows in the surface phase **S**. The only difference from equations (7.16) through (7.21) is that the surface densities $\tilde{\varrho}_\alpha$ and $\tilde{u}$ are used and the pressure is represented by $\tilde{p}_\tau$.

In equation (7.15) the following notations are introduced: $\mathbf{j}_Q^{\Sigma\mathrm{I,II}} = \mathbf{j}'^{\mathrm{I,II}}_H$ are the energy flows through the surfaces Σ_I and Σ_II. These are also the flows $\mathbf{j}_H^{\mathrm{I,II}}$ calculated in coordinate systems moving with velocities $\Delta\mathbf{v}^\mathrm{I} = \mathbf{v}^\mathrm{I} - \mathbf{v}^S$ and $\Delta\mathbf{v}^\mathrm{II} = \mathbf{v}^\mathrm{II} - \mathbf{v}^S$, respectively. Besides, the tensors of the momentum transfer through the boundary surfaces Σ_I and Σ_II are introduced in the form

$$\begin{aligned}\mathbf{j}_p^{\Sigma\mathrm{I,II}}(\mathbf{x}) = \mathbf{T}(\mathbf{x}) - \Big(\frac{\partial\beta^S}{\partial\tilde{u}}\frac{h^{\mathrm{I,II}}}{\beta^{\mathrm{I,II}}} + \sum_\alpha \frac{\partial\beta^S}{\partial\tilde{\varrho}_\alpha}\frac{\varrho_\alpha^{\mathrm{I,II}}}{\beta^{\mathrm{I,II}}}\Big) H(\mathbf{x})\mathbf{U} \\ + \sum_\alpha \Big(\frac{\partial\beta^S}{\partial\tilde{\varrho}_\alpha}\frac{m_\alpha h^{\mathrm{I,II}}}{\beta^{\mathrm{I,II}}} + \sum_{\alpha'} \frac{\partial\nu_\alpha^S}{\partial\tilde{\varrho}_{\alpha'}}\frac{\varrho_{\alpha'}^{\mathrm{I,II}}}{\beta\mathrm{I,II}}\Big) n(\mathbf{x})\mathbf{U}\end{aligned} \tag{7.22}$$

The t_1 dependence of all these flows in equations (7.14) and (7.15) corresponds to the evolution along the dynamic trajectory of the system, for example, $\mathbf{j}_Q(\mathbf{x}, t_1) = \exp\{i\mathcal{L}t_1\}\mathbf{j}_Q(\mathbf{x})$.

It is necessary to distinguish transport processes of different tensor dimension, namely, scalar and tensor processes, in the final form of equations for Δ^1 and Δ^2. In the theory of transport phenomena in an isotropic bulk phase this procedure is well known [7, 4]. A similar procedure can be carried through for the anisotropic surface phase.

We represent the tensor of the momentum flow fluctuations

$$\pi^S = j_p^S - < j_p^S >_q$$

in the form of a sum of the tangential π_τ and normal π_n components: $\pi^S = \pi_\tau + \pi_n$. Here π_τ and π_n are three–dimensional tensors; only tangential components of π_τ are nonzero, and only normal components of π_n are different from zero, such that $\pi_n\mathbf{U}_\tau = \mathbf{U}_\tau\pi_n = 0$.

By analogy, with the dynamic variables of the longitudinal and transverse pressure [15], $\hat{p}_\tau = 1/2\mathrm{Sp}_\tau\mathbf{T}$, $\hat{p}_n = \mathbf{T}_{nn}$, we introduce $\hat{\Pi}_\tau = 1/2\mathrm{Sp}_\tau\pi_\tau$, $\hat{\Pi}_n = \pi_{nn}$. The sign *hat* is used to distinguish molecular (microscopic) pressure from thermodynamic pressure. Then π^S can be represented in the form

$$\pi^S = \mathring{\pi}_\tau + \mathring{\pi}_n + \hat{\Pi}_\tau\mathbf{U}_\tau + \hat{\Pi}_n\mathbf{nn} \tag{7.23}$$

where $\mathring{\pi}_\tau = \pi_\tau - \hat{\Pi}_\tau\mathbf{U}_\tau$, and $\mathring{\pi}_n = \pi_n - \hat{\Pi}_n\mathbf{nn}$ are traceless tensors.

Thus, the total contraction of the tensors $\nabla_\tau \mathbf{v}^S$ and π^S in equation (7.14) takes the form

$$\nabla_\tau \mathbf{v}^S : \pi^S = \hat{\Pi}_\tau \mathrm{div}_\tau \mathbf{v}_\tau^S + (\nabla_\tau \mathring{\mathbf{v}}_\tau^S)^s : \mathring{\pi}_\tau^S + (\nabla_\tau \mathring{\mathbf{v}}_n^S)^s : \mathring{\pi}_n^S \tag{7.24}$$

where $(\nabla_\tau \mathring{\mathbf{v}}_\tau^S)^s$ is the tensor of the rates of tangential shear deformations, and $(\nabla_\tau \mathring{\mathbf{v}}_n^S)^s$ is the tensor of the rates of normal deformations of deflection, and superscript s denotes the symmetric part.

In equation (7.15) it is also necessary to distinguish the tensor and scalar processes. We consider the total contractions $\mathbf{n}\Delta\mathbf{v}^{\mathrm{I,II}} : \pi^{\Sigma\mathrm{I,II}}$, where $\pi^{\Sigma\mathrm{I,II}} = \dot{j}_p^{\Sigma\mathrm{I,II}} - < \dot{j}_p^{\Sigma\mathrm{I,II}} >_q$

It is clear from the structure of these contractions that they contain only normal components of the tensors $\pi^{\Sigma\mathrm{I,II}}$, whose scalar and tensor parts, as for π_n, have the form $\pi_n^{\Sigma\mathrm{I,II}} = \mathring{\pi}_n^{\Sigma\mathrm{I,II}} + \pi_{nn}\mathbf{nn}$, whence

$$\mathbf{n}\Delta\mathbf{v}^{\mathrm{I,II}} : \pi^{\Sigma\mathrm{I,II}} = (\mathbf{n}\Delta\mathring{\mathbf{v}}^{\mathrm{I,II}})^s : (\mathring{\pi})_n^{\Sigma\mathrm{I,II}} + \mathbf{n}\Delta\mathbf{v}^{\mathrm{I,II}} : \pi_{nn}^{\Sigma\mathrm{I,II}} \tag{7.25}$$

These results, in conjunction with equation (7.24), ensure separation of the scalar and tensor longitudinal and transverse processes in the nonequilibrium function of equation (5.3)

Thus, we have completed the construction of the nonequilibrium distribution function taking into account the transport processes in the system.

2.8 Transport Equations for Surface Films

Now we proceed to the construction of equations of the nonideal hydrodynamics of the system under consideration. For this, we calculate the mean flows and forces in the right–hand–side of the conservation laws (5.20), (5.24), and (5.25), taking into account the Curie principle for a system possessing cylindrical symmetry within the surface phase and a spherical symmetry in the bulk phases. Besides, the calculations of all averages are performed in the frame of reference moving with the velocity $\mathbf{v}^S$.

2.8.1 Diffusion

The diffusion flow of the component α is

$$\mathbf{J}_{\alpha}^{S} = <\mathbf{j}_{d\alpha}^{S}(\mathbf{x})(\Delta^1 + \Delta^2)>_q = -\sum_{\beta} D_{\alpha\beta}^{S} \nabla_{\tau} \nu_{\beta}^{S} - D_{\alpha T}^{S} \frac{\nabla_{\tau} T^S}{T^S}$$
$$+\mathbf{n}\sum_{\beta} D_{\alpha\beta}^{\Sigma\mathrm{I}}(\nu_{\beta}^{\mathrm{I}} - \nu_{\beta}^{S}) - \mathbf{n}\sum_{\beta} D_{\alpha\beta}^{\Sigma\mathrm{II}}(\nu_{\beta}^{\mathrm{II}} - \nu_{\beta}^{S}) \tag{8.1}$$
$$-\mathbf{n}D_{\alpha T}^{\Sigma\mathrm{I}}\left(\frac{1}{T^{\mathrm{I}}} - \frac{1}{T^S}\right) + \mathbf{n}D_{\alpha T}^{\Sigma\mathrm{II}}\left(\frac{1}{T^{\mathrm{II}}} - \frac{1}{T^S}\right)$$

The first two terms of this equation correspond to diffusion and thermal diffusion within the surface film, and the last four describe the diffusion and thermal diffusion mass transfer between the surface and bulk phases. For the transport coefficients, the following expressions are obtained (taking into account cylindrical isotropy of the system)

$$\left\{\begin{array}{c} D_{\alpha\beta}^{S} \\ \\ D_{\alpha T}^{S} \end{array}\right\} = \int_{-\infty}^{0} dte^{\varepsilon t} \int d\mathbf{x}' \Theta^S(\mathbf{x}') \left\{\begin{array}{c} (\mathbf{j}_{d\alpha}^{\tau}(\mathbf{x}),\, \mathbf{j}_{d\beta}^{\tau}(\mathbf{x}',t)) \\ \\ (\mathbf{j}_{d\alpha}^{\tau}(\mathbf{x}),\, \mathbf{j}_{Q}^{\tau}(\mathbf{x}',t)) \end{array}\right\} \tag{8.2}$$

$$\left\{\begin{array}{c} D_{\alpha\beta}^{\Sigma\mathrm{I,II}} \\ \\ D_{\alpha T}^{\Sigma\mathrm{I,II}} \end{array}\right\} = \int_{-\infty}^{0} dte^{\varepsilon t} \int_{\Sigma_{\mathrm{I,II}}} ds' \left\{\begin{array}{c} (j_{d\alpha}^{n}(\mathbf{x}),\, j_{d\beta}^{n}(\mathbf{s}',t)) \\ \\ (j_{d\alpha}^{n}(\mathbf{x}),\, j_{Q}^{n}(\mathbf{s}',t)) \end{array}\right\} \tag{8.3}$$

Using the notion:

$$(A(\mathbf{x}), B(\mathbf{x}',t)) =< A(\mathbf{x})(B(\mathbf{x}',t) - < B(\mathbf{x}',t) >_{\mathrm{I}}) >_0 \tag{8.4}$$

it should be noticed that the function ρ_q in the averaging is replaced with the equilibrium Gibbs distribution function ρ_0, equation (1.2.8), corresponding to some average constant values of thermodynamic parameters $\beta_0, \mathbf{v}_0, \nu_0$. Such a procedure is quite justified for a linear approximation in the parameter gradients.

In accordance with the Curie principle all previous transport coefficients are to be scalars, as the phases possess spherical or cylindrical isotropy. It means that correlation functions of the type in equation (8.4) for vectors **A** and **B** may be presented as scalar productions of three- and two-dimensional vectors in the form

$$(\mathbf{A}(\mathbf{x}), \mathbf{B}(\mathbf{x}',t)) = \frac{1}{3} < \mathbf{A}(\mathbf{x}) \cdot (\mathbf{B}(\mathbf{x}',t) - < \mathbf{B}(\mathbf{x}',t) >_0) >_0 \tag{8.5}$$

$$(\mathbf{A}^\tau(\mathbf{x}), \mathbf{B}^\tau(\mathbf{x}', t)) = \frac{1}{2} < \mathbf{A}^\tau(\mathbf{x}) \cdot (\mathbf{B}^\tau(\mathbf{x}', t) - < \mathbf{B}^\tau(\mathbf{x}', t) >_0) >_0 \quad (8.6)$$

2.8.2 Heat flow

equation for the heat flow vector in the surface phase takes the form

$$\begin{aligned} \mathbf{J}_H^S = < \mathbf{j}_H^S(\mathbf{x})(\Delta^1 + \Delta^2) >_q &= -\lambda^S \frac{\nabla_\tau T^S}{T^S} - \sum_\alpha \kappa_\alpha^S \nabla_\tau \nu_\alpha^S \\ &+ \mathbf{n} \sum_\alpha \kappa_\alpha^{\Sigma \mathrm{I}} (\nu_\beta^{\mathrm{I}} - \nu_\beta^S) - \mathbf{n} \sum_\alpha \kappa_\alpha^{\Sigma \mathrm{II}} (\nu_\beta^{\mathrm{II}} - \nu_\beta^S) \\ &- \mathbf{n} \lambda^{\Sigma \mathrm{I}} \Big(\frac{1}{T^{\mathrm{I}}} - \frac{1}{T^S} \Big) + \mathbf{n} \lambda^{\Sigma \mathrm{II}} \Big(\frac{1}{T^{\mathrm{II}}} - \frac{1}{T^S} \Big) \end{aligned} \quad (8.7)$$

The first term with coefficient λ^S is the heat conduction along the surface film, and the terms with coefficients $\lambda^{\Sigma \mathrm{I}}$ and $\lambda^{\Sigma \mathrm{II}}$ describe the heat transfer between the surface phase and the bulk phases **I** and II. The terms with coefficients κ_α^S, $\kappa_\alpha^{\Sigma \mathrm{I}}$, $\kappa_\alpha^{\Sigma \mathrm{II}}$ correspond to the Dufour effect within the surface film and for transfer across the boundary surfaces Σ_{I} and Σ_{II}. For the transport coefficients the following expressions are obtained:

$$\left\{ \begin{array}{c} \lambda^S \\ \\ \kappa_\alpha^S \end{array} \right\} = \int_{-\infty}^{0} dt e^{\varepsilon t} \int d\mathbf{x}' \Theta^S(\mathbf{x}') \left\{ \begin{array}{c} (\mathbf{j}_Q^\tau(\mathbf{x}),\, \mathbf{j}_Q^\tau(\mathbf{x}', t)) \\ \\ (\mathbf{j}_Q^\tau(\mathbf{x}),\, \mathbf{j}_{d\alpha}^\tau(\mathbf{x}', t)) \end{array} \right\} \quad (8.8)$$

$$\left\{ \begin{array}{c} \lambda^{\Sigma \mathrm{I,II}} \\ \\ \kappa_\alpha^{\Sigma \mathrm{I,II}} \end{array} \right\} = \int_{-\infty}^{0} dt e^{\varepsilon t} \int_{\Sigma_{\mathrm{I,II}}} d\mathbf{s}' \left\{ \begin{array}{c} (j_Q^n(\mathbf{x}),\, j_Q^n(\mathbf{s}', t)) \\ \\ (j_Q^n(\mathbf{x}),\, j_{d\alpha}^n(\mathbf{s}', t)) \end{array} \right\} \quad (8.9)$$

The correlation functions on the right–hand–parts of these expressions are defined by equations (8.4) through (8.6).

2.8.3 Momentum flow

To distinguish the scalar and tensor longitudinal and transverse transport processes, we use equation (7.23) to represent the microscopic momentum flow tensor in the anisotropic surface film in the form

$$\mathbf{T}(\mathbf{x}) = \hat{p}_n \mathbf{n}\mathbf{n} + \hat{p}_\tau \mathbf{U}_\tau + \mathring{T}_\tau(\mathbf{x}) + \mathring{T}_n(\mathbf{x}) \quad (8.10)$$

where $\mathring{T}_\tau(\mathbf{x}) = T_\tau(\mathbf{x}) - \mathrm{Sp}_\tau T_\tau(\mathbf{x})/2$, $\mathring{T}_n(\mathbf{x}) = T_n(\mathbf{x}) - \hat{p}_n \mathbf{n}\mathbf{n}$ are the traceless tensors of the normal and tangential components of the momentum flow.

Using the nonequilibrium distribution function, equation (7.13) in conjunction with equations (7.24) and (7.26), we obtain in the comoving frame of reference

$$\sigma^S = -\mathbf{\Pi}^S = -<\mathbf{T}(\mathbf{x})>^S = \sigma_q^S - <\hat{p}_n(\Delta^1+\Delta^2)\mathbf{nn}>_q^S$$
$$-<\hat{p}_\tau \mathbf{U}_\tau(\Delta^1+\Delta^2)>_q^S - <\mathring{\hat{T}}_\tau(\mathbf{x}) + \mathring{\hat{T}}_n(\mathbf{x})(\Delta^1+\Delta^2)>_q^S \tag{8.11}$$

The nonequilibrium stress tensor $\tilde{\sigma}$ results from the integration of equation (8.11) across the layer. Calculating the nonequilibrium terms in equation (8.11) in conjunction with the Curie principle and taking account equation (6.6) for $\tilde{\sigma}_q$, we find

$$\tilde{\sigma} = -\tilde{p}_n\mathbf{U} + \Gamma + (\tilde{\xi}_n\mathbf{nn} + \tilde{\xi}_\tau\mathbf{U}_\tau)\mathrm{div}_\tau\mathbf{v}_\tau^S$$
$$-(\tilde{\xi}_n^{\Sigma_\mathrm{I}}\mathbf{nn} + \tilde{\xi}_\tau^{\Sigma_\mathrm{I}}\mathbf{U}_\tau)\mathbf{n}\cdot\Delta\mathbf{v}^\mathrm{I} + (\tilde{\xi}_n^{\Sigma_\mathrm{II}}\mathbf{nn} + \tilde{\xi}_\tau^{\Sigma_\mathrm{II}}\mathbf{U}_\tau)\mathbf{n}\cdot\Delta\mathbf{v}^\mathrm{II} \tag{8.12}$$
$$+2\tilde{\eta}_\tau[\mathring{\nabla}_\tau\mathbf{v}_\tau^S]^s + 2\tilde{\eta}_n[\mathring{\nabla}_\tau v_n^S]^s - 2\tilde{\zeta}^\mathrm{I}[\mathbf{n}\mathring{\Delta}\mathbf{v}^\mathrm{I}]^s + 2\tilde{\zeta}^\mathrm{II}[\mathbf{n}\mathring{\Delta}\mathbf{v}^\mathrm{II}]^s$$

Here the scalar coefficients

$$\xi_{n,\tau} = \beta^S \int_{-\infty}^{0} dt e^{\varepsilon t} \int d\mathbf{x}'\Theta^S(\mathbf{x}')\left(\hat{p}_{n,\tau}(\mathbf{x}), \hat{\Pi}_\tau(\mathbf{x}',t)\right) \tag{8.13}$$

are the two-dimensional analogs of the coefficient of second (or bulk) viscosity, and they describe the transverse and longitudinal stresses in the system that result from the extension or contraction of the surface film with the rate $\mathrm{div}_\tau\mathbf{v}_\tau^S$. The coefficients

$$\xi_{n,\tau}^{\Sigma_\mathrm{I,II}} = \beta^\mathrm{I,II} \int_{-\infty}^{0} dt e^{\varepsilon t} \int_{\Sigma_\mathrm{I,II}} d\mathbf{s}\Theta^S(\mathbf{s})\left(\hat{p}_{n,\tau}(\mathbf{x}), \hat{\pi}_{nn}^{\Sigma_\mathrm{I,II}}(\mathbf{s},t)\right) \tag{8.14}$$

do not have an analog in the bulk phase. They characterize the viscous stresses associated with the inflow (or outflow) of matter into the surface phase through its boundaries Σ_I and Σ_II with velocity $\Delta v_n^\mathrm{I,II} = v_n^\mathrm{I,II} - v_n^S$.

The tensor coefficients of longitudinal, η_τ, and transverse, η_n, shear viscosity differ from each other. The first describes shear viscosity associated with tangential shear deformations of the film, and the second describes the shear viscosity associated with deflection deformations to the shape of the film. In a local Cartesian coordinate system with z axis along the normal vector $\mathbf{n}$, they can be represented in the form

$$\eta_\tau = \frac{1}{2}\beta^S \int_{-\infty}^{0} dt e^{\varepsilon t} \int d\mathbf{x}' \Theta^S(\mathbf{x}')(\mathring{T}_\tau(\mathbf{x}), \mathring{\pi}_\tau(\mathbf{x}', t)) \tag{8.15}$$

$$\eta_n = \frac{1}{2}\beta^S \int_{-\infty}^{0} dt e^{\varepsilon t} \int d\mathbf{x}' \Theta^S(\mathbf{x}')(\mathring{T}_n(\mathbf{x}), \mathring{\pi}_n(\mathbf{x}', t)) \tag{8.16}$$

The coefficients ζ^{I} and ζ^{II} are characteristic of only heterogeneous systems, and they describe friction in the case of a slip in velocity $\Delta\mathbf{v}_\tau^{\mathrm{I,II}} = \mathbf{v}_\tau^{\mathrm{I,II}} - \mathbf{v}_\tau^S$. In the case of the above–mentioned Cartesian coordinate system, they have the form

$$\zeta^{\mathrm{I,II}} = \beta^{\mathrm{I,II}} \int_{-\infty}^{0} dt e^{\varepsilon t} \int_{\Sigma_{\mathrm{I,II}}} d\mathbf{s}(\mathring{T}_n(\mathbf{x}), \mathring{\pi}_n^{\mathrm{I,II}}(\mathbf{s}, t)) \tag{8.17}$$

The correlation functions of scalars should be considered, for example, as

$$\left(\hat{p}_\tau(\mathbf{x}), \hat{\Pi}_\tau(\mathbf{x}', t)\right) =< \hat{p}_\tau(\mathbf{x})(\hat{\Pi}_\tau(\mathbf{x}', t) - < \hat{\Pi}_\tau(\mathbf{x}', t) >_0) >_0 \tag{8.18}$$

The correlation functions of tensors in equations (8.15) through (8.17) are of special interest.

Owing to cylindrical isotropy, the first of these functions may be represented in the form of the degenerate tensor of rank four as

$$K^\tau_{\alpha\beta\gamma\delta} = (\mathring{T}_{\alpha\beta}, \mathring{\pi}_{\gamma\delta}) = k^\tau \{\frac{1}{2}(\delta_{\alpha\delta}\delta_{\beta\gamma} + \delta_{\alpha\gamma}\delta_{\beta\delta}) - \frac{1}{2}\delta_{\alpha\beta}\delta_{\gamma\delta}\} \tag{8.19}$$

here k^τ is a scalar, $\delta_{\alpha\beta}$ is the Kronecker symbol, $\alpha, \beta, \gamma, \delta = x, y$, where x,and y are local tangential coordinates.

Only eight of the sixteen components of this tensor are not equal to zero:

$$\begin{aligned} K^\tau_{xxxx} &= K^\tau_{yyyy} = \tfrac{1}{2}k^\tau, \quad & K^\tau_{xyxy} &= K^\tau_{yxyx} = \tfrac{1}{2}k^\tau \\ K^\tau_{xxyy} &= K^\tau_{yyxx} = -\tfrac{1}{2}k^\tau, \quad & K^\tau_{xyyx} &= K^\tau_{yxxy} = \tfrac{1}{2}k^\tau \end{aligned} \tag{8.20}$$

Therefore

$$k^\tau = \frac{1}{2}\sum_{\alpha\beta} K^\tau_{\alpha\beta\beta\alpha} = 2K_{xyxy} \tag{8.21}$$

Since the production of the degenerate tensor in the curly brackets in equation (8.19) by the tensor$[\nabla_\tau \overset{\circ}{\mathbf{v}}{}^S_\tau]^s$ is

$$k^\tau \sum_{\gamma\delta} \left\{ \frac{1}{2}(\delta_{\alpha\delta}\delta_{\beta\gamma} + \delta_{\alpha\gamma}\delta_{\beta\delta}) - \frac{1}{2}\delta_{\alpha\beta}\delta_{\gamma\delta} \right\} [\nabla_\tau \overset{\circ}{\mathbf{v}}{}^S_\tau]^s_{\gamma\delta} = k^\tau [\nabla_\tau \overset{\circ}{\mathbf{v}}{}^S_\tau]^s_{\alpha\beta} \quad (8.22)$$

it is possible to substitute in such productions the tensor correlation function K^τ by the scalar value k^τ, or

$$(\overset{\circ}{T}_\tau(\mathbf{x}), \overset{\circ}{\pi}_\tau(\mathbf{x}', t)) \to k^\tau = 2 < T_{xy}(\mathbf{x})(T_{xy}(\mathbf{x}', t)) - < (T_{xy}(\mathbf{x}', t)) >_0) >_0 \quad (8.23)$$

Similarly, the correlation function of the normal components of the tensors in equation (8.16) may be replaced by the scalar value in the form

$$(\overset{\circ}{T}_n(\mathbf{x}), \overset{\circ}{\pi}_n(\mathbf{x}', t)) \to 2 < T_{zy}(\mathbf{x})(T_{zy}(\mathbf{x}', t) - < (T_{zy}(\mathbf{x}', t)) >_0) >_0 \quad (8.24)$$

Thus, the longitudinal shear viscosity coefficient, equation (8.15), of the surface layer can be rewritten using equation (8.23) as

$$\eta_\tau = \beta^S \int_{-\infty}^{0} dt e^{\varepsilon t} \int d\mathbf{x}' \Theta^S(\mathbf{x}') < T_{xy}(\mathbf{x}) T_{xy}(\mathbf{x}', t) >_0 \quad (8.25)$$

and the deflection viscosity coefficient of the surface layer, equation (8.16), can be rewritten using equation (8.24) as

$$\eta_n = \beta^S \int_{-\infty}^{0} dt e^{\varepsilon t} \int d\mathbf{x}' \Theta^S(\mathbf{x}') < T_{zy}(\mathbf{x}) T_{zy}(\mathbf{x}', t) >_0 \quad (8.26)$$

Now we consider the friction coefficients in equation (8.17). The multipliers of the tensors of surface slips are given in equation (8.12):

$$2[\mathbf{n}\Delta\overset{\circ}{\mathbf{v}}{}^{\mathrm{I,II}}]^s = \begin{pmatrix} 0 & 0 & \Delta v_x^{\mathrm{I,II}} \\ 0 & 0 & \Delta v_y^{\mathrm{I,II}} \\ \Delta v_x^{\mathrm{I,II}} & \Delta v_y^{\mathrm{I,II}} & 0 \end{pmatrix} \quad (8.27)$$

Since the system is spherically isotropic in the neighborhood of the boundaries Σ_{I} and Σ_{II}, the correlation functions of the normal components of tensors $\overset{\circ}{T}$ and $\overset{\circ}{\pi}$ are to be of the form

$$(\overset{\circ}{T}_n, \overset{\circ}{\pi}{}_n^{\mathrm{I,II}})_{klmn} = k^{\mathrm{I,II}} \left\{ \frac{1}{2}(\delta_{km}\delta_{\mathrm{ln}} + \delta_{kn}\delta_{lm}) - \frac{1}{3}\delta_{kl}\delta_{mn} \right\} \quad (8.28)$$

where $k^{\mathrm{I,II}}$ are scalars, and $k, l, m, n = x, y, z$. Coefficients $\zeta^{\mathrm{I,II}}$ are written in terms of the particular form of this correlation function:

$$K^{\mathrm{I,II}}_{kl\gamma m} = (\mathring{T}_n, \mathring{\pi}^{\mathrm{I,II}}_n)_{kl\gamma m} k, l = x, y, z; \gamma = x, y; m = z \tag{8.29}$$

It is easy to verify that only four of the components of this tensor are non–zero:

$$K^{\mathrm{I,II}}_{zxxz} = K^{\mathrm{I,II}}_{zyyz} = K^{\mathrm{I,II}}_{xzxz} = K^{\mathrm{I,II}}_{yzyz} = \frac{1}{2} k^{I,\mathrm{II}} \tag{8.30}$$

whence

$$k^{\mathrm{I,II}} = 2 < T_{xz}(\mathbf{x})(T_{xz}(\mathbf{x}', t)) - < (T_{xz}(\mathbf{x}', t)) >_0) >_0 \tag{8.31}$$

or

$$(\mathring{T}_n, \mathring{\pi}^{\mathrm{I,II}}_n) = 2 < T_{xz}(\mathbf{x})(T_{xz}(\mathbf{x}', t)) - < (T_{xz}(\mathbf{x}', t)) >_0) >_0 \tag{8.32}$$

Substituting equation (8.32) into equation (8.17), we obtain the friction coefficients in the final forms

$$\zeta^I = \beta^I \int_{-\infty}^{0} dt e^{\varepsilon t} \int_{\Sigma_I} ds' < T_{xz}(\mathbf{x}) T_{xz}(\mathbf{x}', t) >_0 \tag{8.33}$$

$$\zeta^{\mathrm{II}} = \beta^{\mathrm{II}} \int_{-\infty}^{0} dt e^{\varepsilon t} \int_{\Sigma_{\mathrm{II}}} ds' < T_{xz}(\mathbf{x}) T_{xz}(\mathbf{x}', t) >_0 \tag{8.34}$$

Thus we have derived the final formulae for all transport coefficients from transport equations (8.1), (8.7), and (8.12). Substituting these equations into the conservation laws, equations (5.20), (5.21), (5.24), and (5.25) ensure the final form of the nonideal hydrodynamics equations for a surface system. Combined with wellknown hydrodynamic equations for bulk phases, they provide the full hydrodynamic description of the system with the interphase surface.

Naturally, these formal transport coefficients expressions are not sufficient for solution of any concrete problem. It is necessary to have their numerical values. Presently There are no sufficiently rigorous approaches for calculating the flow correlation functions at interphase boundaries that enter into these expressions. In concrete problems we are thus forced to rely only upon experimental values of relevant transport coefficients. One such problem related to capillary waves will be discussed at the end of the chapter.

2.9 Motion Equations for Excess Quantities

The hydrodynamic equations derived above describe behavior of the true surface densities of the type (2.8). In the equilibrium theory of surface properties, on the other hand, it is common practice to use excess quantities introduced by Gibbs.

According to the Gibbs approach, a dividing surface Σ_o is placed inside the surface layer and an excess density of an extensive property φ is defined as

$$\varphi^* = \int_{-\infty}^{\infty} dz[\varphi(z) - \Theta^o \varphi^{\mathrm{I}} - (1-\Theta^o)\varphi^{\mathrm{II}}] \tag{9.1}$$

where Θ^o is the stepwise Heaviside function, $\Theta^o = 1$ on the left of the surface Σ^o, and $\Theta^o = 0$ on the right of Σ^c; $\varphi(z)$ is the real profile across the layer; φ^{I} and φ^{II} are values of the property φ in the bulk phases in the neighborhood of the surface layer.

In a nonequilibrium system it is also possible [16] to introduce excess quantities similar to equation (9.1). So an excess mass density of α component may be written in the form

$$\varrho_\alpha^*(\mathbf{s},t) = \int_{-\infty}^{\infty} dz\,\left[\varrho_\alpha(\mathbf{x},t) - \Theta^o(\mathbf{x},t)\varrho_\alpha^{\mathrm{I}}(\mathbf{x},t) - (1-\Theta^o(\mathbf{x},t))\varrho_\alpha^{\mathrm{II}}(\mathbf{x},t)\right] \tag{9.2}$$

where a time dependence of the Θ^o-function takes into account motion of the surface Σ^o.

The excess density ϱ_α^* may be written in terms of the true surface density $\tilde{\varrho}_\alpha$:

$$\varrho_\alpha^*(\mathbf{s},t) = \tilde{\varrho}_\alpha(\mathbf{s},t) - \int_{-\Sigma_{\mathrm{I}}}^{\Sigma_{\mathrm{II}}} dz\,\left[\Theta^o(\mathbf{x},t)\varrho_\alpha^{\mathrm{I}}(\mathbf{x},t) + (1-\Theta^o(\mathbf{x},t))\varrho_\alpha^{\mathrm{II}}(\mathbf{x},t)\right] \tag{9.3}$$

It is easy to derive its conservation law, as

$$\begin{aligned}\frac{\partial \varrho_\alpha^*}{\partial t} &= \frac{\partial \tilde{\varrho}_\alpha}{\partial t} - \int_{-\infty}^{\infty} dz \dot{\Theta}^S\left[\Theta^o \varrho_\alpha^{\mathrm{I}} + (1-\Theta^o)\varrho_\alpha^{\mathrm{II}}\right] \\ &\quad - \int_{-\infty}^{\infty} dz \Theta^S\left[\dot{\Theta}^o(\varrho_\alpha^{\mathrm{I}} - \varrho_\alpha^{\mathrm{II}}) + \Theta^o \dot{\varrho}_\alpha^{\mathrm{I}} + (1-\Theta^o)\dot{\varrho}_\alpha^{\mathrm{II}}\right]\end{aligned} \tag{9.4}$$

or

$$
\begin{aligned}
\frac{\partial}{\partial t}\varrho_\alpha^* + \mathbf{v}^S \cdot \nabla_\tau \varrho_\alpha^* &= -\varrho_\alpha^* \mathrm{div}_\tau \mathbf{v} - m_\alpha \mathrm{div}_\tau \mathbf{J}_\alpha^* \\
&-\mathbf{n} \cdot \left(m_\alpha \mathbf{J}_\alpha^{\mathrm{II}}|_{\Sigma_{\mathrm{II}}} - m_\alpha \mathbf{J}_\alpha^{\mathrm{I}}|_{\Sigma_{\mathrm{I}}} \right) - \mathbf{n} \cdot \left(\varrho_\alpha^{\mathrm{II}} \Delta \mathbf{v}^{\mathrm{II}}|_{\Sigma_{\mathrm{II}}} - \varrho_\alpha^{\mathrm{I}} \Delta \mathbf{v}^{\mathrm{I}}|_{\Sigma_{\mathrm{I}}} \right)
\end{aligned} \tag{9.5}
$$

where the conservation law (5.20) of the true surface density and the equations (2.5), (2.6) are used, and similar equations for $\Theta^o(\mathbf{x}, t)$:

$$
\frac{\partial \Theta^o}{\partial \mathbf{x}} = -\mathbf{n}\delta_s^o, \qquad \frac{\partial \Theta^o}{\partial t} = \mathbf{n} \cdot \mathbf{v}\delta_s^o \tag{9.6}
$$

are taken into account.

It is easy to verify that the only distinction between equation (9.5) and the related equation (5.20) is presence of the excess diffusion flow $\mathbf{J}_\alpha^*$ in equation (9.5) substituted for the flow $\tilde{\mathbf{J}}_\alpha^S$ in (5.20).

Similarly, the conservation laws for the momentum and energy excess densities are derived in the forms

$$
\begin{aligned}
\varrho^* \frac{\partial}{\partial t}\mathbf{v}^S + \varrho^* \mathbf{v}^S \cdot \nabla_\tau \mathbf{v}^S &= +\mathrm{Div}_\tau \sigma^* + \mathbf{n} \cdot \left(\sigma^{\mathrm{II}}|_{\Sigma_o} - \sigma^{\mathrm{I}}|_{\Sigma_o} \right) \\
&-\mathbf{n} \cdot \left(\varrho^{\mathrm{II}} \Delta \mathbf{v}^{\mathrm{II}} \Delta \mathbf{v}^{\mathrm{II}}|_{\Sigma_o} - \varrho^{\mathrm{I}} \Delta \mathbf{v}^{\mathrm{I}} \Delta \mathbf{v}^{\mathrm{I}}|_{\Sigma_o} \right)
\end{aligned} \tag{9.7}
$$

$$
\begin{aligned}
\frac{\partial}{\partial t}u^* + \mathbf{v}^S \cdot \nabla_\tau u^* &= -u^* \mathrm{div}_\tau \mathbf{v}^S - \mathrm{div}_\tau J_H^* \\
&+\sigma^* : \nabla_\tau \mathbf{v}^S - \mathbf{n} \cdot (\mathbf{J}_H^{\mathrm{II}}|_{\Sigma_o} - \mathbf{J}_H^{\mathrm{I}}|_{\Sigma_o}) \\
&-\mathbf{n} \cdot \left[\left(u^{\mathrm{II}} - \sigma^{\mathrm{II}} + \frac{1}{2}\varrho^{\mathrm{II}}(\Delta v^{\mathrm{II}})^2 \right) \Delta \mathbf{v}^{\mathrm{II}}|_{\Sigma_o} \right. \\
&\left. - \left(u^{\mathrm{I}} - \sigma^{\mathrm{I}} + \frac{1}{2}\varrho^{\mathrm{I}}(\Delta v^0)^2 \right) \Delta \mathbf{v}^0|_{\Sigma_o} \right]
\end{aligned} \tag{9.8}
$$

where the excess stress tensor σ^* and the excess energy flow $\mathbf{J}_H^*$ are introduced.

All flows are equal to zero in the ideal hydrodynamics approximation, and the excess stress tensor σ_q^* may be represented with the use of equations (6.5) and (9.1) in the form

$$
\sigma_q^* = -p_n^* \mathbf{U} + \gamma^* \mathbf{U}_\tau \tag{9.9}
$$

where

$$
\gamma^* = \gamma \tag{9.10}
$$

as pressures p^{I} and p^{II} of bulk phases are supposed to be isotropic. Notice that in the case of an anisotropy of p^{I} and p^{II}, the surface tension coefficient becomes

$$\gamma^* = \int_{-\infty}^{\infty} dz[p_n(\mathbf{x},t) - p_\tau(\mathbf{x},t) - \Theta^o(p_n^{\mathrm{I}} - p_\tau^{\mathrm{I}}) - (1 - \Theta^o(p_n^{\mathrm{II}} - p_\tau^{\mathrm{II}})] \quad (9.11)$$

Thus the ideal hydrodynamics equations for the excess densities are exactly the same as the relevant equations for the true surface densities.

Transport equations for the excess flows are derived from equations (8.1), (8.7) and (8.11) using the procedure in equation (9.1). As an example we write the final equation for the excess tensor of surface stresses in the form

$$\begin{aligned}\sigma^* &= \int_{-\infty}^{\infty} dz\left[\sigma^S(\mathbf{x},t) - \Theta^o\sigma^{\mathrm{I}} - (1-\Theta^o)\sigma^{\mathrm{II}}\right]\\ &= -p_n^*\mathbf{U} + \Gamma^* + \left(\xi_n^*\mathbf{nn} + \xi_\tau^* U_\tau\right) div_\tau \mathbf{v}_\tau^S\\ &\quad -\left(\tilde{\xi}_n^{\Sigma 0}\mathbf{nn} + \tilde{\xi}_\tau^{\Sigma 0}\mathbf{U}_\tau\right)\mathbf{n}\cdot\Delta\mathbf{v}^{\mathrm{I}}\\ &\quad +\left(\tilde{\xi}_n^{\Sigma 1}\mathbf{nn} + \tilde{\xi}_\tau U_\tau\right)\mathbf{n}\cdot\Delta\mathbf{v}^{\mathrm{II}} + 2\eta_\tau^*\left[\nabla_\tau \mathbf{v}_\tau)\right]^{\overset{\circ}{s}}\\ &\quad +2\eta_n^*\left[\nabla_\tau v_n^S\right]^{\overset{\circ}{s}} - 2\tilde{\zeta}^{\mathrm{I}}\left[\mathbf{n}\Delta\mathbf{v}^{\mathrm{I}}\right]^{\overset{\circ}{s}} + 2\tilde{\zeta}^{\mathrm{II}}\left[\mathbf{n}\Delta v^{\mathrm{II}}\right]^{\overset{\circ}{s}}.\end{aligned} \quad (9.12)$$

The physical meaning of the transport coefficients in this equation is the same as in equation (8.12).

Thus, we conclude that the excess properties and the equations for them do not differ significantly from the relevant true surface densities and equations. In the case of a surface singularity of any property, for example, ϱ_α, the contribution of the integral of the smooth function $\varrho_\alpha^{\mathrm{I}}$ and $\varrho_\alpha^{\mathrm{II}}$ over the macroscopically infinitesimal interval $(\Sigma_{\mathrm{I}}, \Sigma_{\mathrm{II}})$ can be neglected in equation (9.3). As a result we have $\varrho_\alpha^* \simeq \tilde{\varrho}_\alpha$. In the opposite case, where singularity of this property is absent, both ϱ_α^* and $\tilde{\varrho}_\alpha$ are infinitesimal in relation to the bulk densities $\varrho_\alpha^{\mathrm{I}}$ and $\varrho_\alpha^{\mathrm{II}}$ and can be disregarded in the relevant conservation laws, equations (5.20) and (9.5). Then the laws become relations between values of this property and its flow in bulk phases at the interphase boundary.

2.10 The Problem of Boundary Conditions in Hydrodynamics

Here we discuss the situation when there are no singularities of mass or energy in the surface layer. Then the surface flows J_α^S, J_H^S vanish, as does the momentum density, $\tilde{\varrho}v^S$, of the layer. In this case, the whole system of conservation laws for the surface phase degenerates into a system of equations connecting quantities corresponding to the bulk phases I and II. In particular, equation (5.20) or (9.5) becomes

$$\mathbf{n}\cdot\left(J_\alpha^{\mathrm{II}}+\frac{\varrho_\alpha^{\mathrm{II}}}{m_\alpha}\Delta v^{\mathrm{II}}\right)_{\Sigma_0}=\mathbf{n}\cdot\left(J_\alpha^{\mathrm{I}}+\frac{\varrho_\alpha^{\mathrm{I}}}{m_\alpha}\Delta v^{\mathrm{I}}\right)_{\Sigma_0} \tag{10.1}$$

Hereafter we make no difference between Σ_{II} or Σ_{I} and Σ_0, as smooth bulk properties do not change practically along the normal from Σ_{II} or Σ_{I} to Σ_0.

If equation (10.1) is true for all K components of the system ($\alpha = 1, 2, \ldots, K$), then after summing equation (10.1) taking into account the equality $\sum m_\alpha J_\alpha = 0$, it follows that

$$\varrho^{\mathrm{II}}\Delta v^{\mathrm{II}} = \varrho^{\mathrm{I}}\Delta v^{\mathrm{I}} \tag{10.2}$$

Equation (10.2) is the condition of conservation of a normal momentum component in mass transport through an interphase boundary.

Under the same conditions, using equations (5.25) and (10.2), we obtain the generalized nonequilibrium capillary equation

$$\operatorname{Div}\tilde{\sigma}+\mathbf{n}\cdot(\sigma^{\mathrm{II}}-\sigma^{\mathrm{I}})-\mathbf{n}\cdot(\varrho^{\mathrm{II}}\Delta v^{\mathrm{II}}\Delta v^{\mathrm{II}}-\varrho^{\mathrm{I}}\Delta v^{\mathrm{I}}\Delta v^{\mathrm{I}})=0 \tag{10.3}$$

At the interphase boundary of two immiscible liquids $\Delta v_n^{\mathrm{I}} = \Delta v_n^{\mathrm{II}} = 0$, equation (10.3) takes the form

$$\operatorname{Div}\tilde{\sigma}=\mathbf{n}\cdot(\sigma^{\mathrm{I}}-\sigma^{\mathrm{II}}) \tag{10.4}$$

The tangential and normal components of this equation relate transverse and longitudinal gradients of the bulk phase velocities, accounting for surface properties.

The boundary condition for momentum balance at the surface of two immiscible liquids, equation (10.4), is of great interest in hydrodynamics. The general structure of this relation is evident, the basic problem is choosing of a specific form of the surface tension tensor.

Boussinesque [17] was apparently the first to formulate the boundary condition of this type in 1913. Introducing the surface tension tensor by analogy with the bulk situation, he assumed linear dependence on a two–dimensional tensor of deformation rates. Boussinesque assigned two viscosity coefficients to a surface: dilation viscosity (two-dimensional analog of the second (bulk) viscosity) and two-dimensional shear viscosity.

A more elaborate presentation of his theory belongs to Scriven [18]. His results follow from equations (8.12) or (9.12) if we substitute $\zeta_n = \eta_n = 0$.

The most complete phenomenological theory of the surface tension tensor is due to Goodrich [19, 12]. He paid attention to the hypothesis by Shuleikin and Ivanov [20] regarding the viscosity of an oil film on a water surface, related this to a deflection motion of the film, and developed the hypothesis by introducing the three-dimensional surface tension tensor with surface viscosity coefficients. Goodrich used the Gibbs method of a dividing surface and, describing flows of bulk phases with the use of Navier–Stokes equations, obtained the surface viscosity coefficients as excess surface properties. The general form of this surface tension tensor is exactly the same as that of $\tilde{\sigma}$ in equation (9.12) (see Ref. [21]).

To write the tangential and normal components of equation (10.4) in an explicit form, it is necessary to substitute equations (8.12) or (9.12) and related equations for σ^{I} and σ^{II}. For the general case, the formulas obtained are too cumbersome and there is no reason state them. We confine our discussion to the simplest particular cases.

For a homogeneous plane surface film $\operatorname{Div}\tilde{\sigma} = 0$, equation (10.4) translates into the usual boundary condition for the normal components of stress tensors of the bulk phase:

$$\mathbf{n}\cdot\sigma^{\mathrm{I}} = \mathbf{n}\cdot\sigma^{\mathrm{II}}$$

In the ideal hydrodynamics approximation, the boundary condition (10.3) takes the form

$$\nabla\tilde{p}_n + \operatorname{Div}\Gamma + \mathbf{n}\cdot(p^{\mathrm{I}} - p^{\mathrm{II}}) + (\varrho^{\mathrm{I}}\Delta v_n^{\mathrm{I}}\Delta\mathbf{v}^{\mathrm{I}} - \varrho^{\mathrm{II}}\Delta v_n^{\mathrm{II}}\Delta\mathbf{v}^{\mathrm{II}}) = 0 \qquad (10.5)$$

The tangential component of this equation in the case of a homogeneous surface film can be written as

$$\varrho^{\mathrm{I}}\Delta v_n^{\mathrm{I}}(\Delta\mathbf{v}_\tau^{\mathrm{I}} - \Delta\mathbf{v}_\tau^{\mathrm{II}}) = 0 \qquad (10.6)$$

where equation (10.2) is taken into account. It follows from the equation (10.6) that if the normal mass flow $\varrho^{\mathrm{I}}\Delta v_n^{\mathrm{I}}$ is no–zero, the condition of tangential velocity continuity is of the form

$$\Delta v_\tau^{\mathrm{I}} - \Delta v_\tau^{\mathrm{II}} = 0 \quad \text{or} \quad v_\tau^{\mathrm{I}} = v_\tau^{\mathrm{II}} \qquad (10.7)$$

The normal component of equation (10.5) determines the pressure jump over a curved surface as

$$p^{\mathrm{I}} + \varrho^{\mathrm{I}}(\Delta v_n^{\mathrm{I}})^2 = p^{\mathrm{II}} + \varrho^{\mathrm{II}}(\Delta v_n^{\mathrm{II}})^2 + p_\Sigma \tag{10.8}$$

where

$$p_\Sigma = -\mathbf{n} \cdot \operatorname{Div} \Gamma \tag{10.9}$$

is the capillary pressure. Note, that it follows from the definition of Γ, that $\mathbf{n} \cdot \Gamma = 0$ and therefore $\operatorname{div} \mathbf{n} \cdot \Gamma = 0$, hence

$$\mathbf{n} \cdot \operatorname{Div} \Gamma = -(\nabla_\tau \mathbf{n}) \colon \Gamma \tag{10.10}$$

Let the slightly curved boundary surface be described by the relation $z = \zeta(x, y, t)$. The vector normal to this surface has tangential coordinates $-\partial\zeta/\partial x$, $-\partial\zeta/\partial y$. Since the tensor Γ has only two components: $\Gamma_{xx} = \Gamma_{yy} = \gamma$, it is sufficient for calculation of p_Σ to find only two components of the tensor $\nabla_\tau \mathbf{n}$:

$$(\nabla_\tau \mathbf{n})_{xx} = -\frac{\partial^2 \zeta}{\partial x^2} = \frac{1}{R_x}, \quad (\nabla_\tau \mathbf{n})_{yy} = -\frac{\partial^2 \zeta}{\partial y^2} = \frac{1}{R_y} \tag{10.11}$$

where R_x and R_y are curvature radii of the surface $\eta(x, y, t)$. Therefore,

$$p_\Sigma = -\gamma \left(\frac{\partial^2 \zeta}{\partial x^2} + \frac{\partial^2 \zeta}{\partial y^2} \right) = \gamma \left(\frac{1}{R_1} + \frac{1}{R_2} \right) \tag{10.12}$$

where R_1 and R_2 are principal curvature radii $1/R_1 + 1/R_2 = 1/R_x + 1/R_y$.

In the case of two immiscible liquids, equation (10.8) takes the form of the wellknown Laplace formula: $p^{\mathrm{I}} - p^{\mathrm{II}} = \gamma(1/R_1 + 1/R_2)$.

From the conservation law equation (5.26) in the case of the zeroth surface densities of energy $\tilde{u}$ and energy flow $\tilde{J}_H$ the generalized boundary conditions for the bulk energy flows take the form

$$\begin{aligned} &\sigma^* \colon \nabla_\tau \mathbf{v}^S + \mathbf{n} \cdot (\mathbf{J}_H^{\mathrm{I}} - \mathbf{J}_H^{\mathrm{II}}) \\ &+ \left[\left(u^{\mathrm{I}} - \sigma^{\mathrm{I}} + \varrho^{\mathrm{I}} \frac{(\Delta \mathbf{v}^I)^2}{2} \right) \Delta v_n^{\mathrm{I}} - \left(u^{\mathrm{II}} - \sigma^{\mathrm{II}} + \varrho^{\mathrm{II}} \frac{(\Delta \mathbf{v}^{\mathrm{II}})^2}{2} \right) \Delta v_n^{\mathrm{II}} \right] = 0 \end{aligned} \tag{10.13}$$

In the ideal hydrodynamics approximation, taking into account equation (10.2), this boundary condition takes the form

$$w^{\mathrm{I}} + \frac{(\Delta \mathbf{v}^I)^2}{2} = w^{\mathrm{II}} + \frac{(\Delta \mathbf{v}^{\mathrm{II}})^2}{2} - \frac{\gamma}{j} \operatorname{div} \mathbf{v}_\tau^{\mathrm{S}} \tag{10.14}$$

For a flat homogeneous surface layer it becomes

$$w^{\mathrm{I}} + \frac{(\Delta \mathbf{v}^{I})^2}{2} = w^{\mathrm{II}} + \frac{(\Delta \mathbf{v}^{\mathrm{II}})^2}{2} \tag{10.15}$$

Here $w = h/\varrho$ is the enthalpy density relative to a mass unit and $j = \varrho^{\mathrm{I}} \Delta v_n^{\mathrm{I}} = \varrho^{\mathrm{II}} \Delta v_n^{\mathrm{II}}$ is the mass flow through the boundary.

When the expressions obtained in the previous sections for the flows in the surface phase are substituted into the generalized boundary conditions, terms arise with slip or discontinuities of the hydrodynamic variables (velocity, temperature, and chemical potentials) on the boundary of the surface phases. If the values of these discontinuities are not specified, the boundary conditions alone will be insufficient to establish a unique correspondence between the flow in the bulk phases in the neighborhood of the interface. The best known example of this kind is Maxwell's slip boundary condition for a rarefied gas on a wall. Therefore, to illustrate the general approach to the construction of boundary conditions for discontinuities of the parameters, one should consider isothermal shear flow of a gas near a solid surface.

In accordance with the general model of surface phenomena considered in this chapter, we shall assume that the entire gas–solid system consists of three phases: the bulk phases I (gas) and II (solid) and the surface phase **S**. We are not interested in nonequilibrium processes on the surface of the solid itself; therefore, the surface phase S includes only the film of gas adjoining the solid surface. The depth of this film is determined by the roughness of the solid surface (for example, a polished metallic surface it is of order of 10^{-5} cm).

Let us consider the momentum transport in such a film, taking the x axis along the flow of the gas and the z axis along the normal to the surface.

The surface stress tensor equation (8.12) for this system is simplified appreciably, since the surface gas film does not have singular mass or energy densities and does not have surface tension, so that we have

$$-\sigma^{\mathrm{S}} = p^{\mathrm{I}} U + 2\zeta^{\mathrm{I}} [\mathbf{n} \mathring{\Delta} \mathbf{v}^{\mathrm{I}}]^{s} - 2\zeta^{\mathrm{II}} [\mathbf{n} \mathring{\Delta} \mathbf{v}^{\mathrm{II}}]^{s} \tag{10.16}$$

where ζ^{I} and ζ^{II} are the coefficients of friction on the boundaries of the surface phase. Notice that the surface stress tensor is not integrated here over the layer, so we use the superscript S, rather than tilde or an asterisk.

The velocity v_x^S of the surface layer is assumed to be independent of the coordinate z. The value of the velocity may differ both from both the velocity of the bulk gas phase v_x^{I} on the boundary Σ_{I} with the gas film as well as from the velocity $v_x^{\mathrm{II}} = 0$ of the solid surface Σ_{II}. The relative values of discontinuities of the velocities on the surfaces are determined by the condition of equality of the momentum transport through $\Sigma_{\rm I}$ and $\Sigma_{\rm II}$: $\mathbf{n}\cdot\sigma^{\rm S}|_{\Sigma_{\rm I}} = \mathbf{n}\cdot\sigma^{S}|_{\Sigma_{\rm II}}$, where for $\Delta v_x^{\rm I}$ and $\Delta v_x^{\rm II}$ we obtain

$$\zeta^{\rm I}\Delta v_\alpha^{\rm I} = -\zeta^{\rm I}\Delta v_\alpha^{\rm II} \tag{10.17}$$

This equation determines the relationship between the discontinuities of the velocities across the different boundaries of the surface film: the film slips relative to the bulk phases with a velocity at which the forces of friction on its two boundaries are equal.

We now consider the case of an absolutely rough solid surface on which the friction coefficient $\zeta^{\rm II}$ is so large ($\zeta^{\rm II} \gg \zeta^{\rm I}$) that $\Delta v_x^{\rm II} = 0$, i.e., the surface film is completely carried along by the solid surface. The discontinuity of the velocity on the other surface $\Sigma_{\rm I}$ can be determined from the condition of conservation of the momentum on transport through this boundary, $\mathbf{n}\cdot\sigma^{\rm S}|_{\Sigma_{\rm I}} = \mathbf{n}\cdot\sigma^{\rm I}|_{\Sigma_{\rm I}}$, or $\zeta^{\rm I}\Delta v_x^{\rm I} = \eta^{\rm I}\partial v_x^{\rm I}/\partial z$, whence

$$\Delta v_x^{\rm I} = \frac{\eta^{\rm I}}{\zeta^{\rm I}}\frac{\partial v_x^{\rm I}}{\partial z} \tag{10.18}$$

where $\eta^{\rm I}$ is the shear viscosity coefficient of the gas. Thus, the slip is expressed in terms of the ratio of transport coefficients. Note that equation (8.33) for $\zeta^{\rm I}$ differs from the Kubo formula for shear viscosity of a fluid, or

$$\eta^{\rm I} = \beta^{\rm I}\int_{-\infty}^{0} dt\, e^{\varepsilon t}\int dx' < T_{xx}(x)T_{xz}(x',t) >_0 \tag{10.19}$$

only in integration over a surface and not over a volume. Therefore, using the mean-value theorem, we can estimate $\zeta^{\rm I}$ for a rarefied gas as $\zeta^{\rm I} \simeq \eta^{\rm I}/l_0$, where l_0 is the characteristic correlation length in the bulk gas phase, which is the mean free path. Then from the exact (in the framework of this model) formula (10.18) Maxwell's boundary condition follows:

$$\Delta v_x^{\rm I} \simeq l_0\partial v_x^{\rm I}/\partial z \tag{10.20}$$

In the same way one can also consider heat and diffusion slip on the boundaries of the surface film; and again it is not necessary to resort to the traditional methods of kinetic theory, based on the solution of the Boltzmann equation with boundary conditions for the one–particle distribution function.

2.11 Theory of Capillary Waves Damping

Here we consider an application of previous boundary conditions to the problem of capillary waves damping on the surface of a viscous liquid coated with a surface–active substance. The phenomenon has long been known, but until recently theoretical models differed on the relative importance of dissipative processes in bulk phases and a surface phase as such.

It is possible now to analyze this problem in detail using the boundary condition in equation (10.3) following the general work of Levich[22].

Let the plane wave $\zeta(x,t) = \zeta_0 e^{i(kx-\omega t)}$ propagate along the liquid–gas surface $z = 0$. Here $k = æ + i\alpha$, $æ = 2\pi/\lambda$ is the wave number, α is the space damping coefficient, ω is the angular frequency. Assuming the deformation ζ of the plane surface being small and taking into account equations (10.10) through (10.12), we get the tangential and normal components of the boundary condition (10.4) in the forms

$$\frac{\partial\gamma}{\partial x} + (\xi_\tau + \eta_\tau)\frac{\partial^2 v_x}{\partial x^2} = \mu\left(\frac{\partial v_z}{\partial x} + \frac{\partial v_x}{\partial z}\right) \tag{11.1}$$

$$\gamma\frac{\partial^2\zeta}{\partial x^2} + \eta_n\frac{\partial^2 v_z}{\partial x^2} = p^{\mathrm{I}} - p^{\mathrm{II}} + 2\mu\frac{\partial v_z}{\partial z} \tag{11.2}$$

The liquid velocity must satisfy the linearized Navier–Stokes equation

$$\varrho\frac{\partial\mathbf{v}}{\partial t} = -\nabla p + \mu\Delta\mathbf{v} \qquad (\text{here } p = p^{\mathrm{II}}(x,z,t) - p^{\mathrm{I}}) \tag{11.3}$$

This can be seen in the form $\mathbf{v} = \mathbf{v}_1 + \mathbf{v}_2$, $\nabla\times\mathbf{v}_1 = 0$, $\nabla\cdot\mathbf{v}_2 = 0$, which permits the introduction of the potential φ and the stream function ψ as

$$\mathbf{v}_n = \nabla\varphi\,, \quad v_{2x} = -\frac{\partial\psi}{\partial z}\,, \quad v_{2z} = -\frac{\partial\psi}{\partial x} \tag{11.4}$$

From equations (11.3) and (11.4) we get

$$\frac{\partial\psi}{\partial t} = \frac{\mu}{\varrho}\Delta\psi\,, \quad \frac{\partial\varphi}{\partial t} = \frac{1}{\varrho}p\,, \quad \Delta\varphi = 0 \tag{11.5}$$

The solution to this equation system can be found in the form

$$\varphi = Z_1(z)e^{i(kx-\omega t)}\,, \quad \psi = Z_1(z)e^{i(kx-\omega t)} \tag{11.6}$$

Then, taking into account the decay of the functions φ and ψ at $z \to -\infty$, we get from the first and third equations in (11.5):

$$\varphi = ae^{kz}e^{i(kx-\omega t)}\,, \quad \psi = Be^{mz}e^{i(kx-\omega t)} \tag{11.7}$$

where A and B are constant and $m^2 = k^2 - i\omega\varrho/\mu$. The pressure is found from the second equation (11.5) as $p = -i\omega\varrho\varphi$. The kinematic condition for the wave surface $z = \zeta(x,t)$ is in the form

$$\frac{\partial\zeta}{\partial t} = v_z - v_x\frac{\partial\zeta}{\partial x} \approx v_z = -k\varphi - ik\psi\,, \quad \zeta = \frac{k}{i\omega}\varphi + \frac{k}{\omega}\psi \tag{11.8}$$

In the case of a surface-active substance slightly soluble in bulk liquid, we determine the partial derivative $\partial\gamma/\partial x$ with the use of the surface density conservation law (5.21):

$$\frac{\partial \tilde{\varrho}}{\partial t} + v_x \frac{\partial \varrho}{\partial x} = -\tilde{\varrho}\frac{\partial v_x}{\partial x} \tag{11.9}$$

Assuming the solution to this equation in the form $\varrho^* = {\varrho^*}_0 + \varrho_1 e^{i(kx-\omega t)}$ we get

$$\frac{\partial \gamma}{\partial x} = \frac{\partial \gamma}{\partial \varrho^*}\frac{\partial \varrho^*}{\partial x} = -\varepsilon_0 \frac{k^2}{\omega}(kA + imB) \tag{11.10}$$

whence

$$\varrho_1 = -i{\varrho^*}_0 \frac{1}{\omega}(k^2 A + ikmB) \tag{11.11}$$

where $\varepsilon_0 = -\varrho_0 \partial\gamma/\partial\varrho^*$ is the film modulus of surface elasticity. Substituting equations (11.4) and (11.6) through (11.10) into the boundary conditions, equations (11.1) and (11.2), we get

$$k^2\varepsilon(ikA - mB) + \mu\omega[2k^2 A + i(m^2 + k^2)B] = 0 \tag{11.12}$$
$$\sigma k^3(A + iB) - \varrho\omega^2 A - 2\mu k\omega(ikA - mB) = 0 \tag{11.13}$$

where we have introduced the complex modulus of surface elasticity $\varepsilon = \varepsilon_0 - i\omega(\eta_\tau + \xi_\tau)$ and the complex surface tension $\sigma = \gamma - i\omega\eta_n$. The existing condition for nontrivial solutions A and B to system of equations (11.11) and (11.12) gives rise to the dispersion equation

$$(\varrho\omega^2 - \sigma k^3)(\varrho\omega^2 - \varepsilon m k^2) - \varepsilon\sigma k^6 + 4i\varrho\mu\omega^3 k^2 + 4\mu^2\omega^2 k^3(m-k) = 0 \tag{11.14}$$

that determines the complex wave number $k = æ + i\alpha$ as a function of ω.

The dispersion equation (11.13) differs from the Levich [22] result by the presence of the imaginary parts $i\omega(\eta_\tau + \xi_t au)$ and $i\omega\eta_n$ of the complex modulus of surface elasticity ε and the complex surface tension, respectively. Introducing the dimensionless variables

$$S = \frac{1}{\sigma}(\frac{\omega\mu^3}{\varrho})^{1/2}, \quad Q = \frac{k}{(\omega\varrho/\mu)^{1/2}}, \quad b = \frac{\varepsilon}{\sigma} \tag{11.15}$$

equation (11.13) becomes

$$bQ^6 + 4S^2Q^4 + SQ^3 - 4iS^2Q^2 - S^2 = (bQ^5 + 4S^2Q^3 - bSQ^2)\sqrt{Q^2 - i} \tag{11.16}$$

After squaring, we get the dispersion equation in the final form

$$
\begin{aligned}
& ib^2Q^{10} + 2b(1+b)SQ^9 + (8(1+b)S^3 - 2ib^2S)Q^7 \\
& +(S^2 - 2bS^2 - b^2S^2 - 16iS^4)Q^6 - 8i(1+b)S^3Q^5 \\
& +(ib^2S^2 - 24S^4)Q^4 - 2S^3Q^3 + 8iS^4Q^2 + S^4 = 0
\end{aligned}
\tag{11.17}
$$

Thus, the dispersion equation (11.13) has been reduced to a polynomial of degree 10 in Q, with the complex coefficients. This equation can be solved with a computer. Fifteen years ago the solution was difficult, but now it is a routine procedure. We consider only those roots that satisfy the following conditions: 1) they are o be roots of the starting equation (11.15); 2) $Re\sqrt{Q^2 - i} > 0$, as a consequence of condition $Re\, m > 0$, providing a decay of φ and ψ as $z \to -\infty$ 3) $Re\, Q > 0$ and $Im\, Q > 0$, that provide a decay of waves as $x \to \infty$. The final results are demonstrated in Figure 2.1. It is seen that the decay coefficient rises significantly with increasing deflection viscosity coefficient η_n.

Similar computations on the basis of the Goodrich theory were performed by Earnshaw [23] for the particular case of the pure water surface where $\varepsilon_0 = 0$. He found that the capillary wave decay was very sensitive to increasing η_n and weakly depended on other surface viscosity coefficients.

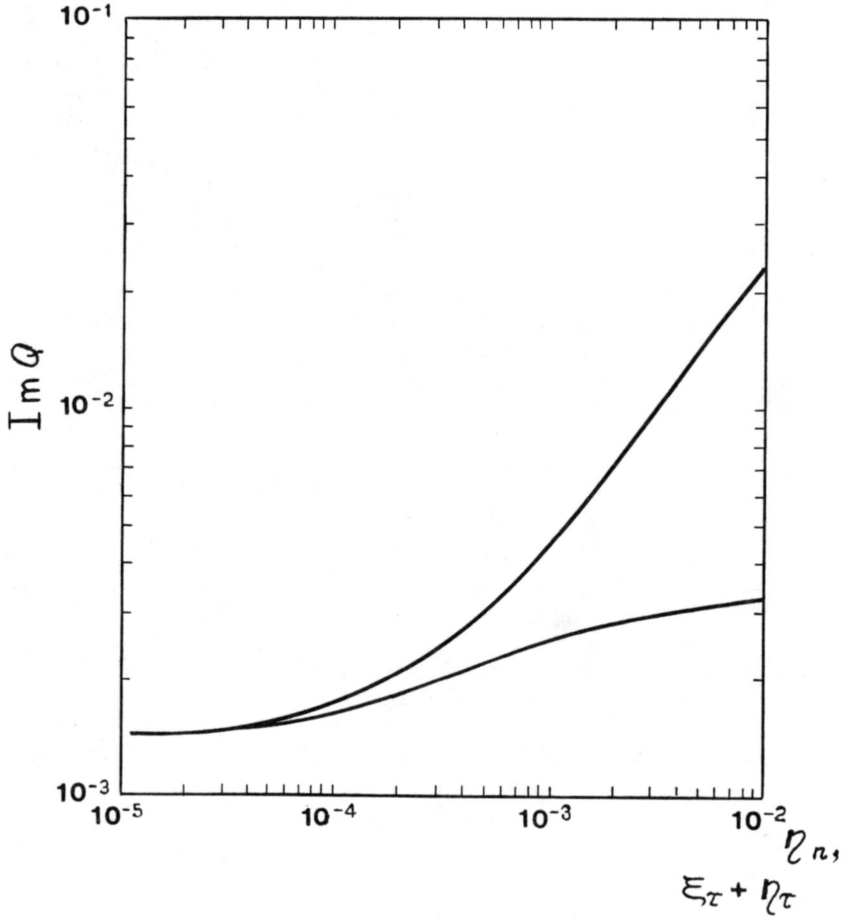

Figure 2.1: Decay coefficient, Im Q, as a function of the deflection viscosity η_n (upper branch) and tangential surface viscosities, $(\xi_\tau + \eta_\tau)$ (lower branch) for $\gamma = 70\,dyn/cm$, $\omega = 2\pi \cdot 10^3$.

Chapter 3

Shock Waves in Gases as Interface Surfaces

3.1 Introduction

A gas–dynamical discontinuity constitutes a surface dividing two gas areas of significantly different values for some physical properties. As an example, a shock front divides the upstream gas flow of relatively low temperature and density from the downstream flow, of high temperature and density.

In this respect, a shock wave is similar to an interphase surface. Moreover, as it was at first noticed by the author and G. A. Korol'kov, a shock front possesses a surface tension, further underscoring their generic character.

One can study a shock wave in a gas using the Boltzmann equation for one–particle distribution function, but not by using the Liouville equation of the previous chapter, when studying nonequilibrium processes on the interphase boundaries, e.g., liquid–liquid, liquid–gas, etc. This fact facilitates the task and permits us to obtain explicitly profiles of all gas–dynamical parameters in the shock layer. However, there is a new problem caused by a strong nonequilibrium state of the gas. This is why there is no satisfactory theory for a shock wave, even for the simplest case of the monatomic ideal gas (with the ideal gas equation of state and nonvanishing viscosity and heat conductivity).

The parameters used for a hydrodynamic description of fluid flows are time–dependent local values for the main thermodynamic properties, namely, density, pressure, temperature, entropy etc. It is common practice to use the same set of parameters to describe flows possessing discontinuous surfaces, but this seems inadequate because in equilibrium thermodynamics of a surface system we must supply an additional thermodynamic parameter; the surface tension coefficient, γ.

In Chapter 2 surface tension has been associated with pressure anisotropy in the surface layer; the coefficient γ was defined by equation (2.6.7). Such a formula is well known in equilibrium theory of surface systems, but its immediate application to a discontinuity in an ideal gas may prove problematic for a reader accustomed to the traditional explanation of the surface tension by intermolecular attractions. To get rid of such a stereotypet, when present, a purely thermodynamic definition of γ is developed in Appendix B without resorting to molecular properties.

On the other hand, a gas–dynamical discontinuity may be realized only in a strongly nonequilibrium gas flow.Its surface tension is provided by an anisotropy of the temperature in the shock layer in contrast to an interphase surface, as an example of which, in a gas–liquid, the surface tension is due to an anisotropy of the mean virial of the intermolecular force.

As all the known derivations of boundary conditions of gas flow discontinuity have overlooked the surface tension and related capillary pressure we shall rederive them in Section 3.2 using the previously described method of a dividing surface developed by Gibbs especially to derive boundary conditions at interphase surfaces with allowance for intrinsic surface properties. Essencially the desired relations have already been formulated in the preceding chapter; here they are represented [24] as a generalization of well–known Rankine–Hugoniot shock jump relations allowing for the pressure anisotropy within a shock layer.

The surface tension coefficient of the shock wave will be estimated in Sec. 3.3 using hydrodynamic theory of a viscous shock layer and kinetic theory of a shock wave structure. The values of γ obtained permit comparison of relative contributions of the surface pressure, that is, capillary pressure behind the shock front and the viscous tension in the downstream flow. It appears that in the case of a weak shock wave the contributions are of the same order. In the case of a strong shock wave, the capillary pressure is one or two orders of magnitude greater than the viscous tension.

Clearly, such a significant property can appear to be of first importance in certain applications. It may be assumed that the shock wave capillary pressure has significant influence on the expansion dynamics of spherical or cylindrical shock waves with small curvature radii.

As an example of such an application, we consider shock wave stability relative to corrugation–like disturbances of the shock front.

This problem is analyzed allowing for viscous stresses in the downstream flow and surface tension of the shock front.

In Section 3.4, a set of equations is derived for small fluctuations of the flow parameters generated by a small corrugation–like disturbance in a plane shock wave front.

A characteristic equation for this set will be analyzed and new stability conditions will be derived in Section 3.5. These conditions are compared with other predictions and experimental data in Section 3.6. It is appears that the viscous stresses and the capillary pressure have significant impact on shock wave stability, resulting in an absolute instability of the shock wave in the regime of spontaneous sound emission, as it has been seen experimentally.

3.2 The Jump Conditions on Curved Discontinuities

We begin with continuous description of a gas flow containing a shock wave. Such a description implies that all gas properties vary continuously through the shock layer from upstream to downstream flow values.

The system of equations for mass, momentum, and energy conservation, including the gas flow and the shock layer, is

$$\begin{aligned} &\frac{\partial \varrho}{\partial t} + \operatorname{div} \varrho \mathbf{v} = 0 \\ &\frac{\partial}{\partial t}\varrho \mathbf{v} + \operatorname{Div}(\mathbf{\Pi} + \varrho \mathbf{v}\mathbf{v}) = 0 \\ &\frac{\partial}{\partial t}\left(u + \frac{\varrho v^2}{2}\right) + \operatorname{div}\left[\left(u + \frac{\varrho v^2}{2}\right)\mathbf{U} + \mathbf{\Pi}\right]\cdot \mathbf{v} = -\operatorname{div}\mathbf{q} \end{aligned} \tag{2.1}$$

The upstream and downstream gas flows are considered here as the bulk phases, I and II. In accordance with the standard notations in gas dynamics, the relevant gas parameters will be denoted hereafter by the superscripts 0 and 1, respectively.

To describe the flow we take into account the heat flux $\mathbf{q}(\mathbf{x}, t)$ and the viscous stresses

$$\sigma'(\mathbf{x}, t) = -\mathbf{\Pi}(\mathbf{x}, t) + p(\mathbf{x}, t)\mathbf{U}$$

In the gas flow upstream and downstream of the shock front the density $\varrho(\mathbf{x}, t)$, velocity $\mathbf{v}(\mathbf{x}, t)$, pressure $p(\mathbf{x}, t)$, viscous stress tensor $\sigma'(\mathbf{x}, t)$, internal energy density $u(\mathbf{x}, t)$, and heat flux $\mathbf{q}(\mathbf{x}, t)$, become ϱ_0, v_0, p_0, σ'_0, u_0, $\mathbf{q}_0$, and ϱ_1, v_1, p_1, σ'_1, u_1, $\mathbf{q}_1$, respectively.

For the general case, these properties were defined in Chapter 2, but for the particular case of gas flow in which the interparticle interactions are neglected, they can be redefined as moments of a one–particle velocity distribution function, $f(\mathbf{c}, \mathbf{x}, t)$, obeying the Boltzmann equation. Then we have

$$\varrho(\mathbf{x}, t) = < m >_f \,, \qquad \mathbf{v}(\mathbf{x}, t) = \frac{< \mathbf{c} >_f}{< m >_f}$$

$$\mathbf{\Pi} = p\mathbf{U} - \sigma' = m < (\mathbf{c} - \mathbf{v})(\mathbf{c} - \mathbf{v}) >_f \tag{2.2}$$

$$\mathbf{q} = m < (\mathbf{c} - \mathbf{v})(\mathbf{c} - \mathbf{v})^2 >_f$$

Within the shock layer, the profiles of these quantities can be obtained using either the Navier–Stokes approximation (for a weak shock wave), or the methods of kinetic gas theory. Outside the shock layer, profiles are described by the equations of hydrodynamics.

Since the shock wave structure will not be of intrinsic interest henceforth, the gas flow of two different scales can be approximated by considering the shock layer as a surface of a discontinuity.

To derive jump conditions accounting for the surface properties of the discontinuity, we use the Gibbs approach discussed in Section 2.9 of Chapter 2.

Introducing the z axis, directed flow–wise along the normal $\mathbf{n}$ to the dividing surface $\Sigma(t)$, located within the moving shock layer and acting by the operation (2.9.1) on equation (2.1), we get the set

$$\frac{\partial \varrho^*}{\partial t} + \mathrm{div}_\tau(\varrho \mathbf{v}_\tau)^* = (v_{0n} - v_\Sigma)\varrho_0 - (v_{1n} - v_\Sigma)\varrho_1 - (\varrho v_n)^*\Big(\frac{1}{R_1} + \frac{1}{R_2}\Big)$$

$$\begin{aligned} &\frac{\partial}{\partial t}(\varrho \mathbf{v})^* + \mathrm{Div}_\tau(\varrho \mathbf{v}\mathbf{v} - \sigma')^* \\ &\quad = \mathbf{n} \cdot [(p_0 - p_1)\mathbf{U} + \sigma'_1] + p_0 \mathbf{v}_0 (v_{0n} - v_\Sigma) - p_1 \mathbf{v}_1 (v_{1n} - v_\Sigma) \\ &\quad\quad - \frac{\mathbf{n}}{R_1}(\varrho v_n^2 - \varrho v_{x1}^2)^* - \frac{\mathbf{n}}{R_2}(\varrho v_n^2 - \varrho v_{x2}^2)^* \\ &\quad\quad + \frac{\mathbf{n}}{R_1}(\sigma'_{nn} - \sigma'_{11})^* + \frac{\mathbf{n}}{R_2}(\sigma'_{nn} - \sigma'_{22})^* \end{aligned}$$

$$\begin{aligned} &\frac{\partial}{\partial t}\left(u + \frac{\varrho v^2}{2}\right)^* + \mathrm{div}_\tau \mathbf{J}_\tau^* = \left(u_0 + \frac{\varrho_0 v_0^2}{2}\right)(v_{0n} - v_\Sigma) + p_0 v_{0n} \\ &\quad - \left(u_1 + \frac{\varrho_1 v_1^2}{2}\right)(v_{1n} - v_\Sigma) - p_1 v_{1n} + \sigma'_{1nn} v_{1n} - q_{1n} - J_n^*\Big(\frac{1}{R_1} + \frac{1}{R_2}\Big) \end{aligned}$$

$$\mathbf{J} = [(u + p + \varrho v^2/2)\mathbf{U} - \sigma'] \cdot \mathbf{v} + \mathbf{q}\,, \qquad v_\Sigma = \dot{\Sigma} \tag{2.3}$$

Here the wellknown expressions were used for the divergences of a vector and a tensor in curvilinear coordinates z, x_1, x_2 pertaining to the surface

Σ; τ is the tangential direction and $\mathbf{U}$ is the unit tensor. The surface divergences and gradients are defined in Appendix A.

The radii of curvature, R_1 and R_2, are given in terms of Lame coefficients $H_z = 1$, H_1, and H_2:

$$\frac{1}{R_1} = \frac{1}{H_1}\frac{\partial H_1}{\partial z}, \qquad \frac{1}{R_2} = \frac{1}{H_2}\frac{\partial H_2}{\partial z}.$$

There are two terms with excess viscous stresses on the right hand side of the second equation of the set (2.3). These terms form the surface pressure p_Σ, which can be written as

$$p_\Sigma = (1/R_1 + 1/R_2)\gamma \tag{2.4}$$

Here the surface tension coefficient γ is defined as

$$\gamma = -(\sigma'_{nn} - \sigma'_{\tau\tau})^* = -\int_{-\infty}^{\infty} (\sigma'_{nn} - \sigma'_{\tau\tau})\,dz \tag{2.5}$$

for the zero order of the shock front curvature, where $\sigma'_{nn}(z)$ and $\sigma'_{\tau\tau}(z)$ are viscous stresses in the plane shock layer.

It should be noted that the surface tension coefficient γ is independent of a dividing surface position, as opposed to all other excess values in equation (2.3).

In the model of a shock wave as a discontinuity surface without accumulation, or redistribution of excess densities of mass, momentum, or energy, we are to consider only those excess parameters that are independent of dividing surface position. In our case the coefficient γ is the only one of this kind.

Thus, we obtain the following extended Rankine–Hugoniot shock jump relations for the curved discontinuity:

$$\begin{aligned}
&(v_{0n} - v_\Sigma)\varrho_0 = (v_{1n} - v_\Sigma)\varrho_1 \\
&p_0 + \varrho_0 v_{0n}(v_{0n} - v_\Sigma) \\
&\qquad = p_1 + \varrho_1 v_{1n}(v_{1n} - v_\Sigma) - \sigma'_{1nn} + p_\Sigma \\
&\varrho_0 v_{0\tau}(v_{0n} - v_\Sigma) = \varrho_1 v_{1\tau}(v_{1n} - v_\Sigma) - \sigma'_{1n\tau} \\
&(u_0 + \varrho_0 v_0^2/2)(v_{0n} - v_\Sigma) + p_0 v_{0n} \\
&\qquad = (u_1 + \varrho_1 v_1^2/2)(v_{1n} - v_\Sigma) + p_1 v_{1n} - \sigma'_{1nn} v_{1n} + q_{1n}
\end{aligned} \tag{2.6}$$

The resulting boundary conditions differ from the wellknown shock jump relations in a viscous and heat conducting gas by the introduction of the additional term for the capillary pressure, p_Σ. Estimations of this term will be performed in the next sections.

Notice that in essence we have rederived the boundary conditions presented in Section 2.10 of Chapter 2. This has been done intentionally, keeping in mind their importance in gas dynamics. This is why we used the wellknown gasdynamics conservation laws (2.1) as a starting point.

3.3 Estimation of the Surface Tension Coefficient

Substituting the Newtonian expression for viscous stresses into equation (2.5), the surface tension coefficient in the Navier–Stokes approximation becomes

$$\gamma_{\rm NS} = -2\int_{-\infty}^{+\infty} \mu(z)\frac{\partial v_n}{\partial z} dz \tag{3.1}$$

The viscosity coefficient $\mu(z)$ is supposed to be dependent on the coordinate z through the temperature dependence in the form $\mu(z) \propto [T(z)]^\alpha$ where the exponent α varies from 0.5 to 1.2, depending on the molecular interaction potential. In the viscous shock layer approximation, the functions $T(z)$ and $v_n(z)$ are hyperbolic tangents, if there are no relaxation processes. Then

$$\begin{aligned}
\gamma_{\rm NS} &= \frac{2}{1+\alpha}(v_0 - v_1)\frac{(T_1/T_0)^{\alpha+1} - 1}{(T_1/T_0) - 1}\mu_0 \\
\gamma_{\rm NS}(\mathrm{M}_0 \gg 1) &\simeq [2/(1+\alpha)](v_0 - v_1)\mu_1 \\
\gamma_{\rm NS}(\mathrm{M}_0 \simeq 1) &\simeq 2(v_0 - v_1)\mu_0
\end{aligned} \tag{3.2}$$

It is known that in a strong shock wave with the relaxation processes the temperature increases initially from T_0 to a plateau value T^0, and then decreases down in the relaxation zone to a downstream value T_1 . A profile of this kind can be approximated in the kinetic zone as

$$T(z) = \frac{T^0 + T_0}{2} + \frac{T^0 - T_0}{2}\tanh\frac{2z}{\delta_0}$$

and, in the relaxation zone, by

$$T(z) = \frac{T_1 + T^0}{2} + \frac{T_1 - T^0}{2}\tanh\frac{2z}{\delta_1}$$

Using the same method, we can write a two-step profile of the gas velocity decreasing from v_0 to v^0, and then from v^0 to v_1, where $v_0 > v^0 > v_1$.

On combining these two–step profiles for $T(z)$ and $v(z)$, we find from equation (3.1)

$$\gamma_{\rm NS} \simeq \frac{2}{1+\alpha}(v_0 - v_1)\mu(T^0) \tag{3.3}$$

Comparison estimations for the capillary pressure p_Σ and the viscous stress σ'_{1nn} in the downstream flow are obtained from equations (2.4) and (2.5), taking into account equation (3.3):

$$\frac{\sigma'_{1nn}}{\varrho_1 v_1^2} \simeq \frac{\mu_1}{\varrho_1 v_1^2}\frac{\partial v_{1n}}{\partial z} \simeq \frac{\mu_1}{\varrho_1 v_1}\left(\frac{1}{R_1} + \frac{1}{R_2}\right) = \varepsilon_R$$

$$\frac{p_\Sigma}{\varrho_1 v_1^2} \simeq \left(\frac{T^0}{T_1}\right)^\alpha \beta\varepsilon_R\,, \qquad \beta = \frac{v_0}{v_1}$$

Here ε_R is the inverse Reynolds number, relative to the curvature radius R as characteristic length; $1/R = 1/R_1 + 1/R_2$.

For the shock wave in argon ($\mathrm{M}_0 = 30$, $p_0 = 1\,\mathrm{Torr}$, $T_0 = 300\,K$) where ionization relaxation takes place, Chubb [25] has found $\beta \simeq 20$ and $T^0/T_1 \simeq 6$; hence p_Σ may be two orders of magnitude greater than σ'_{1nn}.

A similar estimation can be obtained for a gas with an oscillatory–rotational relaxation. According to the kinetic gas theory of a shock wave in nitrogen [26], we have $\beta \simeq 15$ and $T^0/T_1 \simeq 3$ at $\mathrm{M}_0 = 20$; p_Σ is then one order of magnitude greater than σ'_{1nn}.

It then follows that one should take into account p_Σ when writing the shock jump conditions to within ε_R.

The above estimation of γ is obtained within the framework of the hydrodynamic theory of shock wave structure. This theory is known to describe the shock wave structure satisfactorily only within the limits of low Mach numbers ($\mathrm{M}_0 \simeq 1$).It is the opposite case, for a large M_0, that is of practical interest in the problem of shock wave instability. However, γ is an integral parameter of the layer, so we can expect it to be described satisfactorily with the approximate hydrodynamic theory. To verify this, the coefficient (2.5) will be calculated for a monatomic gas, and the result will be compared with that obtained from kinetic gas theory.

For a monatomic gas of rigid spheres of diameter d and mass m the viscosity is

$$\mu_0 = \frac{5\pi}{32}\lambda_0\varrho_0\left(\frac{8kT_0}{m}\right)^{1/2}, \quad \lambda_0 = \frac{m}{\sqrt{2}\pi d^2\varrho_0}$$

Then, using the wellknown expressions for T_1/T_0 and v_1/v_0 in terms of M_0, we obtain from equation (3.2)

$$\gamma_{\mathrm{NS}} = \frac{25}{96}\left(\frac{3\pi}{10}\right)^{1/2} \lambda_0 p_0 \frac{(5\,\mathrm{M_0}^4 + 14\,\mathrm{M_0}^2 - 3)^{3/2} - 64\,\mathrm{M_0}^3}{\mathrm{M_0}^2(5\,\mathrm{M_0}^2 + 3)} . \tag{3.4}$$

For small and large M_0, this expression becomes, respectively,

$$\gamma_{\mathrm{NS}}(\mathrm{M}_0 \simeq 1) = \frac{25}{8}(3\pi/10)^{1/2}(\mathrm{M}_0 - 1)\lambda_0 p_0 \tag{3.5}$$

$$\gamma_{\mathrm{NS}}(\mathrm{M}_0 \gg 1) = 0.57\lambda_0 p_0\,\mathrm{M_0}^2 \tag{3.6}$$

To find γ within the kinetic gas theory, we use the wellknown Mott–Smith theory of stationary plane shock wave structure. In this formulation the molecular velocity distribution function is assumed to be the sum of two Maxwellians, namely

$$f(\mathbf{c}, z) = f_0(\mathbf{c}, z) + f_1(\mathbf{c}, z)$$

$$f_\alpha(\mathbf{c}, z) = n_\alpha(z)(m/2\pi kT_\alpha)^{3/2} \exp[-m(\mathbf{c} - \mathbf{v}_\alpha)^2/2kT_\alpha]\,, \quad \alpha = 0, 1$$

where T_α and $\mathbf{v}_\alpha$ are assumed to be constant, while the particle number densities are found to be

$$n_0(z) = \frac{n_0}{2}\left(1 - \tanh\frac{2z}{\delta_0}\right), \qquad n_1(z) = \frac{n_1}{2}\left(1 + \tanh\frac{2z}{\delta_0}\right)$$

There are many modifications of the original Mott–Smith theory, leading to different values of the shock–wave thickness δ_0. As the limit $\mathrm{M}_0 \to 1$, the theories give different gas viscosities [27]. The only bimodal theory that gives the correct value of the monatomic gas viscosity μ_0 is the original Mott–Smith theory.

According to this theory the shock–wave thickness δ_0 is given by

$$\delta_0 = \lambda_0 \left[\frac{6}{5}\left(\frac{5}{3\pi}\right)^{1/2} \frac{(\mathrm{M_0}^2 - 1)(5\,\mathrm{M_0}^4 + 30\,\mathrm{M_0}^2 - 3)^{1/2}}{(5\,\mathrm{M_0}^2 + 3)(\mathrm{M_0}^2 + 3)} \psi(\mathrm{M}_0)\right]^{-1}$$

$$\psi(\mathrm{M}_0) = 1 + \frac{D}{7} - \frac{D^2}{126} + \frac{D^3}{1386} - \ldots$$

$$D = \frac{15(\mathrm{M_0}^2 - 1)^2}{2(5\,\mathrm{M_0}^4 + 30\,\mathrm{M_0}^2 - 3)}$$

and λ_0 is the mean free path in the upstream flow.

Using the distribution function (3.6) the mean flow velocity and the pressure tensor are found from equations (2.2):

$$\begin{aligned}
v(z) &= < c_z > /[n_0(z) + n_1(z)] \\
&= [n_0(z)v_0 + n_1(z)v_1]/[n_0(z) + n_1(z)] \\
\sigma_{\tau\tau}(z) &= m < c_x^2 >= n_0(z)kT_0 + n_1(z)kT_1 \\
\sigma_{nn}(z) &= m < [c_z - v(z)]^2 > \\
&= \sigma_{\tau\tau}(z) + mn_0(z)n_i(z)/[n_0(z) + n_1(z)] \cdot (v_0 - v_1)^2.
\end{aligned}$$

Substituting $\sigma'_{nn} = \sigma_{nn} - p$ and $\sigma'_{\tau\tau} = \sigma_{\tau\tau} - p$ into equation (2.5) and integrating over the shock layer yields the surface tension coefficient in the Mott–Smith approximation,

$$\gamma_{\text{MS}} = \frac{5}{4}\lambda_0 p_0 B^{-1}(\text{M}_0{}^2 - 1)\ln(v_0/v_1) \tag{3.7}$$

Using equation (3.7), one obtains

$$\gamma_{\text{MS}}(\text{M}_0 \simeq 1) = \frac{25}{8}(3\pi/10)^{1/2}(\text{M}_0 - 1)\lambda_0 p_0 \tag{3.8}$$

$$\gamma_{\text{MS}}(\text{M}_0 \gg 1) \simeq 0.92\lambda_0 p_0 \, \text{M}_0{}^2 \tag{3.9}$$

Note that equation (3.8) is exactly the same as equation (3.5) while the value in equation (3.9) exceeds the value in equation (3.6) by a factor of two or less. Therefore, the Navier–Stokes theory is believed to describe correctly the order of magnitude for γ, since the Mott–Smith theory agrees well with the experimental data on the shock wave structure.

From equations (3.4) and (3.7) we obtain $T_0 = 300\,\text{K}$, $\gamma \simeq 10^3$ dynes/cm for $\text{M}_0 = 10$. The numerical integration of Bird's results [28] for $\sigma_{\tau\tau}$ and σ_{nn} gives $\gamma \simeq 10^3$ dynes/cm at $T_0 = 300\,\text{K}$ and $\text{M}_0 = 8$. The similar analysis of measured $\sigma_{\tau\tau}$ and σ_{nn} in the shock wave in helium [29] at $T_0 = 160\,\text{K}$ and $\text{M}_0 = 1.59$ gives $\gamma \simeq 20$ dynes/cm. These values of γ agree well with equations (3.5) and (3.8).

Such a large surface tension of the shock wave naturally affects its properties (for a comparison, the surface tension of water has $\gamma = 70$ dynes/cm under normal conditions). This surface tension effect may explain the unexpectedly large pressure behind the front of a spherical shock wave observed in the experiments [30].

The remainder of this chapter is devoted to another application where the surface tension is presumed to be of great importance.

3.4 Equations for Small Disturbances

Let us consider a plane stationary shock wave of arbitrary intensity in a viscous gas with no thermal conductivity. We prescribe a weak corrugation-like disturbance for the shock front as $\zeta(x,t) = \zeta_0 \exp[i(kx - \omega t)]$, so that $v_\Sigma = -i\omega\zeta$. This disturbance generates harmonic fluctuations of entropy δs, pressure δp, velocity δv, and specific volume δV, proportional to $\exp[i(kx + lz - \omega t)]$ in the flow behind the shock front $z > 0$.

The downstream gas flow can be treated as an isentropic one in the linear approximation on disturbances, because deviations from isentropicity are of second order in velocity disturbances.

Thus we have from equation (2.1) the following set of linearized equations

$$(vl - \omega)\delta s = 0 \tag{4.1}$$

$$[vl - \omega - i\nu(\frac{4}{3}l^2 + k^2)]\delta v_z - \frac{1}{3}i\nu kl\delta v_x + Vl\delta p = 0 \tag{4.2}$$

$$[vl - \omega - i\nu(l^2 + \frac{4}{3}k^2)]\delta v_x - \frac{1}{3}i\nu kl\delta v_z + Vk\delta p = 0 \tag{4.3}$$

$$(vl - \omega)\delta p + (1/V)c^2(k\delta v_x + l\delta v_z) = 0 \tag{4.4}$$

It follows from equation (4.1) that two types of disturbances exist. The first one is called an entropy–vortex disturbance which has $\delta s_e \neq 0$, $vl_e - \omega = 0$. Substituting $l = l_e = \omega/v$ into the equations (4.2) through (4.4) gives

$$-i\nu(\frac{4}{3}l_e^2 + k^2)\delta v_{ez} - \frac{1}{3}i\nu l_e\delta v_{ex} + Vl_e\delta p_e = 0 \tag{4.5}$$

$$-i\nu(l_e^2 + \frac{4}{3}k^2)\delta v_{exz} - \frac{1}{3}i\nu l_e\delta v_{ez} + Vk\delta p_e = 0 \tag{4.6}$$

$$k\delta v_{ex} + l_e\delta v_{ez} = 0 \tag{4.7}$$

After multiplying equation (4.5) by k and equation (4.6) by l_e and summing the results, we have

$$-\frac{4}{3}i\nu(k^2 + l_e^2)(k\delta v_{ex} + l_e\delta v_{ez}) + (k^2 + l_e^2)V\delta p_e = 0$$

From this equation and equation (4.7) it follows that $\delta p_e = 0$. The second type of disturbance is called acoustic and has $\delta s_a = 0$, $vl_a - \omega \neq 0$. For this latter disturbance equations (4.2) through (4.4) become

$$[vl_a - \omega - i\nu(\frac{4}{3}l_a^2 + k^2)]\delta v_{az} - \frac{1}{3}i\nu kl_a\delta v_{ax} + Vl_a\delta p_a = 0 \tag{4.8}$$

$$[vl_a - \omega - i\nu(l_a^2 + \frac{4}{3}k^2)]\delta v_{ax} - \frac{1}{3}i\nu k l_a \delta v_{az} + V l_a \delta p_a = 0 \tag{4.9}$$

$$(vl_a - \omega)\delta p_a + (1/V)c^2(k\delta v_{ax} + l_a \delta v_{az}) = 0 \tag{4.10}$$

After multiplying equation (4.8) by k and equation (4.9) by l_a and deducing the second result from the first one, we have

$$[vl_a - \omega - i\nu(l_a^2 + k^2)](k\delta v_{az} - l_a \delta v_{ax}) = 0$$

From this

$$k\delta v_{az} - l_a \delta v_{ax} = 0 \tag{4.11}$$

Now equations (4.9) and (4.10), respectively, become

$$[vl_a - \omega - \frac{4}{3}(k^2 + l_a^2)]\delta v_{ax} + Vk\delta p_a = 0 \tag{4.12}$$

$$(vl_a - \omega)\delta p_a + (1/Vk)c^2(k^2 + l_a^2)\delta v_{ax} = 0 \tag{4.13}$$

The consistency condition for these two homogeneous equations gives a dispersion equation for acoustic disturbances in a moving viscous gas.

$$\left(\tfrac{l_a}{k} - \tfrac{\omega}{kv}\right)^2 - \tfrac{4}{3}i\varepsilon\left(1 + \tfrac{l_a^2}{k^2}\right)\left(\tfrac{l_a}{k} - \tfrac{\omega}{kv}\right) = \tfrac{1}{\mathrm{M}_1^2}\left(1 + \tfrac{l_a^2}{k^2}\right) \tag{4.14}$$

$$\varepsilon = \tfrac{\nu k}{v}\,.$$

This dispersion equation allows the use of one of two equations, (4.12) or (4.13).

Equations (4.7), (4.11), and (4.12) are supplemented with the boundary conditions (1.6), which account for the flow shear viscosity and surface tension of the shock wave. In linear approximation, these boundary conditions take the following form:

$$\left(1 - i\nu\frac{\omega}{v^2}\right)\delta v_{ex} + \left(1 - 2i\nu\frac{l_a}{v}\right)\delta v_{ax} - i\nu\frac{k}{v}\delta v_{ez} - ik(v_0 - v)\zeta = 0 \tag{4.15}$$

$$\begin{aligned}\frac{1}{3}i\nu\frac{k}{v}\delta_{ax} + \left(1 - i\nu\frac{\omega}{v^2}\right)\delta v_{ez} + \left(1 - \frac{2}{3}i\nu\frac{l_a}{v}\right)\delta v_{az}\\ -\frac{j}{2}\delta V + \frac{1}{2j}\delta p_a + \left(\frac{1}{3}\nu + \frac{1}{2j}\gamma\right)k^2\zeta = 0\end{aligned} \tag{4.16}$$

$$\begin{aligned}2i\nu\frac{k}{v}\delta v_{ex} + \frac{2}{3}i\nu\frac{k}{v}\delta v_{ax} - \frac{4}{3}i\nu\frac{l_a}{v}\delta v_{az} + j\delta V + \frac{1}{j}\delta p_a\\ -\left(2i\frac{\omega}{jv_0}(p - p_0) - \frac{2}{3}\nu k^2 - \frac{1}{j}\gamma k^2\right)\zeta = 0\end{aligned} \tag{4.17}$$

$$
\begin{aligned}
-2i\nu\frac{k}{v}\Psi\delta v_{ex} &+\frac{2}{3}i\nu\frac{k}{v}\Psi\left(2\frac{l_a^2}{k^2}-1\right)\delta v_{ax}-\frac{1}{j}\left(\frac{\partial p}{\partial V}\right)_H\delta V \\
&+\frac{1}{j}\delta p_a-\left(\frac{2}{3}\nu\Psi-\frac{1}{j}\Phi\gamma\right)k\zeta^2=0
\end{aligned}
\tag{4.18}
$$

$$
\begin{aligned}
\Phi &= -\frac{V+V_0}{2\partial w/\partial p-(V-V_0)} \\
\Psi &= -\frac{V_0-V}{V_0+V}\Phi, \qquad j=\frac{v}{V}
\end{aligned}
\tag{4.19}
$$

and

$$
\left(\frac{\partial p}{\partial V}\right)_H=-\left(\frac{\partial w_1}{\partial V_1}-\frac{1}{2}(p_1-p_0)\right)\left(\frac{\partial w_1}{\partial p_1}-\frac{1}{2}(V_1+V_0)\right)^{-1} \tag{4.20}
$$

is the derivative of the Rankine–Hugoniot shock adiabat.

Thus we received a homogeneous system of seven equations (4.7), (4.11), (4.12), and (4.15) through (4.18) for the seven variables ζ, δp_a, δv_{ax}, δv_{az}, δv_{ex}, δv_{ez}, and δV.

A necessary and sufficient condition for this homogeneous system to have a nontrivial solution is for the characteristic determinant, composed of the coefficients of its equations, to vanish. As a result of a slightly difficult calculation of this seventh-order determinant within the linear approximation in small parameters ε and $\Gamma=\beta\gamma k/(p-p_0)$, we derive a characteristic equation (we now omit the subscript 1):

$$
\begin{aligned}
&\left(\beta+\frac{\omega^2}{k^2v^2}+i\Gamma\frac{\omega}{kv}\right)\left(\frac{l_a}{k}-\frac{\omega}{kv}\right)(1+\varphi) \\
&+\left(1+\frac{\omega^2}{k^2v^2}\right)\left(2\frac{\omega}{kv}+i\Gamma(\Phi\varphi+1)\right)-i\varepsilon N=0
\end{aligned}
\tag{4.21}
$$

where

$$
\begin{aligned}
\varphi &= j^2 \left(\frac{\partial V}{\partial p}\right)_H \\
N = &\left(\frac{\omega}{kv}\right)^4 (1-\varphi) + \left(\frac{\omega}{kv}\right)^3 \frac{l_a}{k}(1+\varphi) - \left(\frac{\omega}{vk}\right)^2 \\
&\times \left[1 + 2\beta\varphi(1+\Psi) - \frac{4}{3}\left(\frac{l_a}{k}\right)^2 \varphi(1+\Psi) - \frac{4}{3}\varphi\Psi - \frac{7}{3}\varphi\right] \\
&- \frac{\omega}{kv}\frac{l_a}{k}[\varphi - 3 - 2\beta(\varphi+1)] \\
&+ \frac{4}{3}\beta\left(\frac{l_a}{k}\right)^2 \varphi(1+\Psi) + \frac{2}{3}\beta\varphi(2-\Psi) \\
&+ 2\beta + \frac{2}{3}\frac{\omega}{kv}\frac{v_0}{v_0 - v}\left(\frac{\omega}{kv}\frac{l_a}{k}\right)\varphi(1+\Psi) \\
&- \frac{2}{3}\frac{v_0}{v_0 - v}\left(1 + \frac{\omega}{kv}\frac{l_a}{k}\right)(1-\varphi\Psi)
\end{aligned}
\tag{4.22}
$$

The characteristic equation (4.21) is treated together with the acoustic dispersion equation (4.14). This system of equations differs from that of Dyakov only in the terms containing Γ and ε, which account for viscous stresses within the shock layer and in gas flow behind the shock, respectively.

3.5 Shock Wave Stability Analysis

Dyakov [31]made use of linearized ideal hydrodynamics equations for gas flows upstream and downstream of the shock. Also, he used the Rankine–Hugoniot jump relations for the shock wave, treated as a discontinuous surface. According to Dyakov, the plane shock wave may be absolutely stable, absolutely unstable, or neutrally stable with respect to corrugation–like surface disturbances, depending on the slope of the Hugoniot shock adiabat. Later, the boundary of the neutrally stable region was defined by Iordanskii [32] and Kontorovitz [33].

Swan and Fowles [34] have analyzed the Dyakov theory and, established its consistency with subsequent results.

However, it was shown elsewhere experimentally that the region of neutrally stable waves in terms of Dyakov–Iordanskii–Kontorovitz appeared to be one of absolutely unstable waves.

A proposed approach for interpretation of this experimental phenomenon was to dispense with the hydrodynamical model of shock waves and investigate relaxational processes in the shock wave in more detail; however, this approach has yet to meet with success.

Fowles [35] confirmed the Iordanskii–Kontorovitz boundaries of the region of acoustic wave emission and proposed that real gas effects might be responsible for the instability observed in [36].

A further solution to this problem was proposed [37] as a consequence of an influence of the light emission by relaxing gas in the shock layer on the upstream gas flow. It was found that the neutrally stable regime shock wave became unstable, due to a partial ionization of the upstream flow. However, such an explanation did not account for a similar instability observed in CO_2, at relatively low values for the Mach number, where ionization was absent.

The present approach to this problem is to include dissipative processes into the hydrodynamical model of the shock wave. It will be shown here, that the inclusion of viscous stresses in the shock layer (the surface tension) and downstream gas flow does not shift the boundaries of stability regions; instead, this results in the above–mentioned absolute instability in the "neutral stability region".

In the frame of linearized theory, the loss of shock wave stability is associated with a solution that increases exponentially with time, and decreases with z. So, the basic conditions for instability are as follows:

$$\operatorname{Im}\omega > 0\,, \qquad \operatorname{Im} l_a > 0 \tag{5.1}$$

Hence, the shock wave stability problem reduces to that of determining a range of φ values, for which solutions $\omega(k,\varphi)$ and $l_a(k,\varphi)$ of equations (4.14) and (4.21) satisfy conditions (5.1).

For dimensionless variables $g+ih=-i\omega/kv$ and $t+iu=il_a/k$, the set of equations (4.14) and (4.21) becomes

$$\begin{aligned}(1-g^2-2igh+h^2)&[2ig-2h+i\Gamma(\Phi\varphi+1)]\\ &=(-g^2-2igh+h^2+\beta-\Gamma g-i\Gamma h)\\ &\quad\times(ig-h+it-u)(1+\varphi)+i\varepsilon N\end{aligned} \tag{5.2}$$

$$\begin{aligned}(ig-h+it-u)^2&+\frac{4}{3}i\varepsilon(1-t^2-2iut+u^2)\\ &\times(ig-h+it-u)=(1-t^2-2iut+u^2)/\,\mathrm{M}_1^2\end{aligned} \tag{5.3}$$

First, we shall find the threshold values of φ corresponding to the stability boundary $\operatorname{Im}\omega = 0$, $\operatorname{Im} l_a > 0$, and then analyze the shock wave stability when φ values are beyond these threshold values.

Substitution of $g = \operatorname{Im}(\omega/kv)$ into equations (5.2) and (5.3) produces the following set of equations for their imaginary and real parts:

$$t(\beta + h^2)(1 + \varphi) = \Gamma[(\Phi\varphi + 1)(1 + h^2) - h(u + h)(1 + \varphi)] - \varepsilon \operatorname{Re} N \tag{5.4}$$

$$2h(1 + h^2) = (\beta + h^2)(h + u)(1 + \varphi) - \Gamma ht(1 + \varphi) + \varepsilon \operatorname{Im} N \tag{5.5}$$

$$\mathrm{M}_1^2(u + h)^2 - t^2 = (1 + u^2 - t^2)(1 + \frac{4}{3}\varepsilon \mathrm{M}_1^2 t) \tag{5.6}$$

$$\mathrm{M}_1^2(u + h)[1 + \frac{2}{3}(\varepsilon/t)(1 + u^2 - t^2)] = 0 \tag{5.7}$$

Now we have, from equation (5.4),

$$\begin{aligned} t = \{\ &\Gamma[(\Phi\varphi + 1)(1 + h^2) - h(u + h)(1 + \varphi)] \\ &-\varepsilon \operatorname{Re} N\}/(\beta + h^2)(1 + \varphi) \end{aligned} \tag{5.8}$$

Let us first consider the nonwave solution $h = 0$, $u = 0$ complying with equations (5.5) through (5.7). In this case, we obtain a cubic from equation (5.6) equation for t, with the small parameter ε by t^3. The asymptotic solution for the cubic equation is

$$t_1 = -[3(1 - \mathrm{M}_1)/4\varepsilon \mathrm{M}_1^2] + [2/3(1 - \mathrm{M}_1^2)] \tag{5.9}$$

From equations (5.8) and (5.9) for $h = u = 0$, the lower threshold value is

$$\varphi_1 = -1 + O(\varepsilon^2) \tag{5.10}$$

This threshold value coincides with the corresponding Dyakov result up to ε^2.

For the wave solution, $h \neq 0$, it is possible to omit terms with $\Gamma t \propto \varepsilon t \propto t^2 \propto \varepsilon^2$, and the terms with $\varepsilon \operatorname{Im} N$ in equations (5.5) through (5.7), since from the reality of coefficients for polynomial N (4.22) and condition $\operatorname{Im}\omega = 0$, it follows that $\operatorname{Im} N \propto t \propto \varepsilon$.

When the flow viscosity is neglected ($\varepsilon = 0$), equations (5.6) and (5.7) give

$$h_0 = -[(1 - \mathrm{M}_1^2)^{1/2} / \mathrm{M}_1] + O(\Gamma^2)$$
$$u_0 = -[\mathrm{M}_1 / (1 - \mathrm{M}_1^2)^{1/2}] + O(\Gamma^2) \tag{5.11}$$

and substitution of equation (5.11) into equation (5.5) gives

$$\varphi_0 = [(1 - \mathrm{M}_1^2 - \beta \mathrm{M}_1^2)/(1 - \mathrm{M}_1^2 + \beta \mathrm{M}_1^2)] + O(\Gamma^2) \tag{5.12}$$

This threshold value agrees with that of Iordanskii and Kontorovitz, up to terms of the second order of Γ. The value φ_0 determines the boundary of the mode of the sound spontaneous emission.

Consider now the effect of the gas flow viscosity on this boundary. As $\operatorname{Re} N(h_0, u_0, \varphi_0) \propto h_0^4 \propto \mathrm{M}_1^4$, then $\xi = 2\varepsilon/3t \propto -\mathrm{M}_1^2$, $|\xi| \ll 1$. From equations (5.6) and (5.7), we obtain, for the lowest order of ξ,

$$h = h_0 + O(\xi^2)\,, \qquad u = u_0 + O(\xi^2)$$

Substitution of these expressions into (5.5) gives the same threshold value φ_0 as in equation (5.12) within ξ^2.

Thus, including viscous stresses in the shock layer and gas flow behind the shock front does not influence the boundary of the mode of sound spontaneous emission.

To prove an absolute instability of a shock wave in spontaneous sound emission mode, we set $\varphi = \varphi_0 + \Delta$, and analyze equations (5.2) and (5.3) in the linear approximation on Δ, taking into account $\operatorname{Im}\omega \neq 0$ ($g \neq 0$). We then obtain the following linear inhomogeneous system of equations, rather then equations (5.4) through (5.7):

$$a_i h' + b_i t' + c_i u' + d_i g = e_i \Delta\,, \quad i = 1,2,3,4$$

Here, the coefficients a_i, b_i, c_i, d_i and e_i are expressed in the threshold solutions h_0, u_0, t_0, and φ_0 of equations (5.4) through (5.7), the primed values being deviations from these solutions.

Within ε^2 and ξ^2, the solution of this system for $\operatorname{Im}\omega$ is given by

$$\operatorname{Im}\omega = -kvg = -[(1 - \mathrm{M}_1^2)/\,\mathrm{M}_1^2(1 + \varphi_0)]t_0 kv\Delta \tag{5.13}$$

From this, as soon as the second condition in (5.1) is fulfilled, the first condition ($\operatorname{Im}\omega > 0$) is also fulfilled, provided that $\Delta > 0$, i.e., $\varphi > \varphi_0$.

The two terms in the numerator of expression (5.8) for t_0 may be of opposite signs in the region of the stability threshold $\varphi \simeq_0$.

Indeed, according to Griffiths et al. [36], $\varphi_0 \ll 1$ for argon and CO_2. Therefore. the first term including the surface tension (i.e., the viscous stresses within the shock layer) becomes

$$\Gamma\varphi(1+h^2)(\Phi-1) \simeq \frac{\Gamma}{\mathrm{M}_1^2} j^2 \frac{\partial w/\partial p}{\partial w/\partial V - (p-p_0)/2} \tag{5.14}$$

From the definition (4.20) for $(\partial p/\partial V)_H$, it is clear that with increasing v_0, both the numerator and denominator on the right-hand side of equation (5.14), positive for small v_0, remain positive along the forward run of the shock adiabat, with $(\partial p/\partial V)_H$ being continuous and negative for this run. A change in the sign of $(\partial p/\partial V)_H$ associated with the local extremum of the shock adiabat causes a similar change in sign of the denominator of equation (5.14). In this case, the surface tension destabilizes the shock. The denominator remains positive even though $(\partial p/\partial V)_H$ changes its sign while passing through $\pm\infty$. For this point of the adiabat, $\partial w/\partial p = (V_0 + V)/2$, thus the numerator and the whole expression (5.14) remain positive in the vicinity of this point. One may thus argue that the surface tension stabilizes the shock wave in this case.

It follows from equation (4.22) that the second term in the numerator of equation (5.8) is negative, as could be expected from general physical considerations. In other words, the flow viscosity should strengthen the spatial damping of the acoustic perturbations, i.e., it should give a negative contribution to t_0. According to equation (5.13), it means that the flow viscosity destabilizes the shock front. So it may happen that the surface tension and the viscous stresses in downstream flow produce competitive effects, that is, surface tension stabilizes the shock wave and the viscous stresses destabilize it. If we use a rough approximation of equation (4.22) for $\varphi_0 \ll 1$, the second term in the numerator of (5.8) becomes $\varepsilon\,\mathrm{Re}\,N \simeq -\varepsilon h_0^4 \simeq -\varepsilon\,\mathrm{M}_1^{-4}$. In contrast to equation (5.14) there is the multiplier M_1^{-4} in this term. As M_1 is far less than unity, the destabilizing influence of the gas flow viscosity can surpass the stabilizing influence of the surface tension. Therefore, the absolute instability is caused by the viscous stresses.

A number of experiments were carried out to detect instability of the plane shock wave in argon and CO_2, see Figure 3.1. Solid, crossed, and open symbols show categories “unstable”, “doubtful”, and “stable” shocks, respectively. Vertical line symbols show cases in which the disturbance is attributed to the contact surface. Different shapes of the symbols correspond to different operation types. (Figure from [36] with permission of IOP Publishing Ltd.)

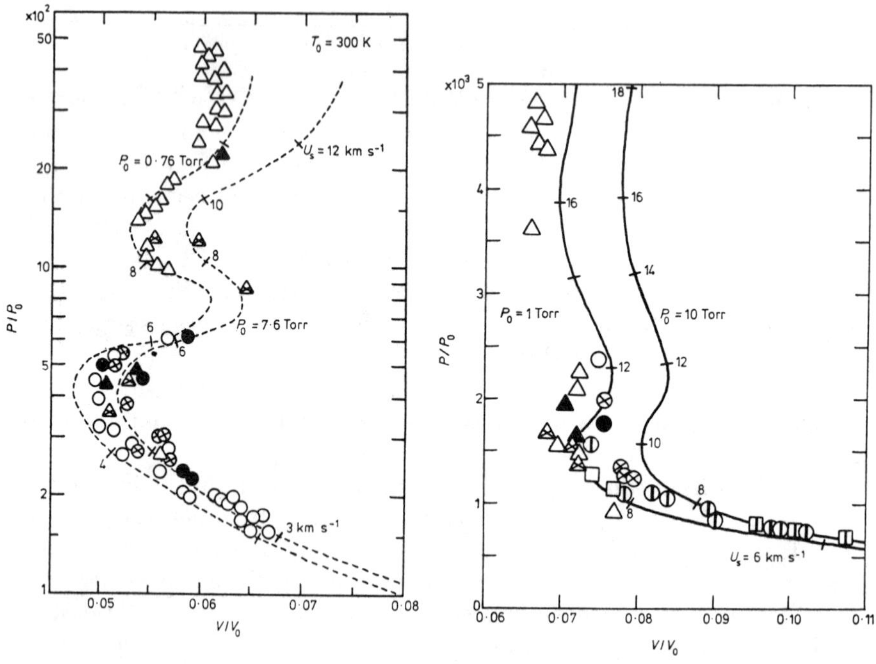

Figure 3.1: Experimental stability results for shock waves in CO_2 (left) and argon (right) superimposed on the computed equilibrium Hugoniot curves.

All the cases of instability occured when $\varphi > \varphi_0$ on "backward runs" of shock adiabat, adjacent to the vertical runs and on the "forward run"; see Figure 3.2. (Figure from [36] with permission of IOP Publishing Ltd.)

The backward runs correspond to $v_0 > 5\,\text{km/sec}$ for CO_2 and $v_0 > 10\,\text{km/sec}$ for argon, while the forward run instability region for CO_2 corresponds to $v_0 \simeq 3.8\,\text{km/sec}$. For each case the instability conditions (5.1) hold, so the present theory agrees quite well with the experiments by Griffiths et al.

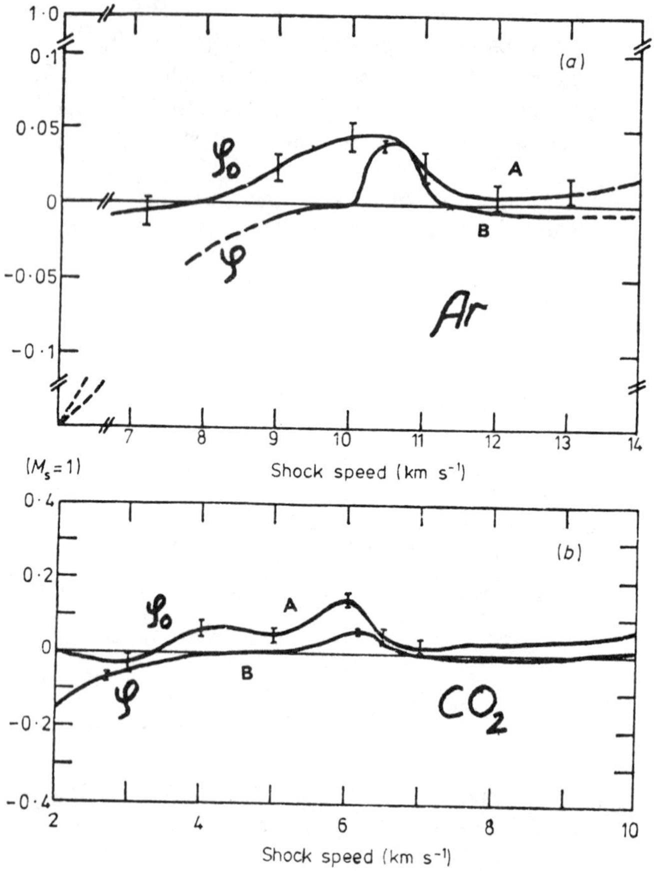

Figure 3.2: The lower boundary φ_0, equation (5.12), of the sound emission mode (curve A) and the term φ, equation (4.22), (curve B) for (a) argon and (b) carbon dioxide shock waves. (a) $p_0 = 1.0\,\text{Torr}$, $T_0 = 293\,K$; (b) $p_0 = 7.6\,\text{Torr}$, $T_0 = 300\,K$.

Chapter 4

Nucleation Theory: Fluctuations of Diffusion Flow across the Interphase Boundary

4.1 Introduction

This chapter is devoted to the study of the diffusion flow fluctuations on the liquid–gas interphase boundary as illustrated by the nucleation phenomenon, that is, growth of a liquid embryo from a gas–vapor mixture. Indeed, at an initial stage of an embryo growth, the thermodynamic potential of the system increases up to some critical value. Hence, this process is prohibited from a purely thermodynamic standpoint; nevertheless it does occure. To explain the nucleation process, it is necessary to take into account fluctuations of the diffusion flow across the embryo surface.

The classical phenomenological nucleation theory developed by Gibbs, Volmer, Becker, Doering, and Zeldovich has been discussed in Frenkel's monograph [38] and treats an embryo growth as a random Markov process. A Fokker–Planck type equation for the size distribution function has been derived. It is known as the Becker equation or (Becker–Zeldovich equation) and has the form

$$\frac{\partial f(g,t)}{\partial t} = \frac{\partial}{\partial g}\left\{D(g)\left[\frac{\partial \Delta\Phi(g)}{\partial g} f(g,t) + kT\frac{\partial f(g,t)}{\partial g}\right]\right\} \qquad (1.1)$$

where g is the embryo size (the number of molecules); $D(g)$ is the nucleation coefficient, and

$$\Delta\Phi(g) = (\mu^{\mathrm{I}} - \mu^{\mathrm{II}})g + \gamma A(g) \tag{1.2}$$

is the thermodynamic potential of embryo formation from vapor with temperature T and chemical potential μ^{II}. Here μ^{I} is the chemical potential of a bulk liquid phase, γ is the surface tension coefficient, and $A(g)$ is the surface area of the embryo.

Here, on the basis of the general approach developed in Chapter 2, a new statistical–mechanical approach [39] to nucleation theory is discussed. The nucleation process is considered to be a result of diffusion–flow fluctuations across the embryo–vapor interphase surface. This new approach allows for a change in vapor concentration in the vicinity of an embryo and applies for arbitrary Knudsen numbers. Sections 4.2–4.5 will be devoted to this formalism.

For the particular case of a free molecular regime where diffusion processes in the gas mixture are completely neglected, another approach to the nucleation theory was developed by the author [40] with the use of a formal quantum–mechanical description of vapor molecules transfers from the gas into the embryo and vice versa. Using this description, regular methods of nonequilibrium statistical mechanics result in the derivation of the Becker equation, including effects due to latent heat. We discuss this theory in Sections 4.6–4.8.

4.2 Total Nonequilibrium Distribution Function

We shall investigate a system consisting of a nonequilibrium subsystem—a liquid embryo interacting with a gaseous mixture. Henceforth, this embryo is supposed to be large enough to be considered a macroscopic subsystem. At first we assume also that the entiresystem is isothermal, i.e., the temperature of the embryo and the gas are the same and constant.

The assumption that the nucleation process is isothermal means that the processes of heat exchange between the growing embryo and the thermostat occur much more rapidly than its growth. It follows from this assumption, in particular, that the distribution function for all degrees of freedom (including collective degrees of freedom) of the embryo remains an equilibrium function at all instants of time. Thus, the only nonequilibrium processes in the system are the growth of the embryo and its attendant diffusion flow in the gas mixture to (or from) the embryo.

The total Hamiltonian of the system is

$$\mathcal{H} = \sum_{\alpha i} \left[\frac{p_{\alpha i}^2}{2m_i} + \frac{1}{2} \sum_{\beta j} U^{\alpha\beta}(\mathbf{r}_{\alpha i} - \mathbf{r}_{\beta j}) \right] \tag{2.1}$$

where $\mathbf{r}_{\alpha i}$ and $\mathbf{p}_{\alpha i}$ are, respectively, the coordinate and momentum of the ith molecule of the α type; the subscripts α, and β are equal to 1 for vapor molecules and equal to 2 for gas–carrier molecules; $U_{\alpha\beta}$ is a binary potential of intermolecular interaction.

To formally describe a space separation of the complete system into two subsystems (the embryo and the gas mixture) it is convenient to use the one–step Heaviside function $\Theta(\mathbf{x}, t)$, nonvanishing (and equal to one) only for $\mathbf{x}$, outside the embryo volume. This function is directly related to the Θ^o–function introduced in Section 2.9 of Chapter 2 through the equation $\Theta^o = 1 - \Theta$. The derivatives are

$$\nabla\Theta(\mathbf{x}, t) = \mathbf{n}\delta_S(\mathbf{x}, t)\,, \quad \frac{\partial}{\partial t}\Theta(\mathbf{x}, t) = -\mathbf{n}\mathbf{v}^S\delta_S(\mathbf{x}, t) \tag{2.2}$$

where δ_S is the surface δ function, $\mathbf{n}$ is the normal to the embryo surface, and $\mathbf{v}^S$ is the velocity of this surface motion.

We introduce microscopic densities of particle numbers in the system in the form

$$n_\alpha(\mathbf{x}) = \sum_i \delta(\mathbf{x} - \mathbf{r}_{\alpha i}) \tag{2.3}$$

and a dynamic density in the space of embryo size in the form

$$n(g) = \delta(g - N^{\mathrm{I}}) \tag{2.4}$$

where

$$N^{\mathrm{I}} = \int [1 - \Theta(\mathbf{x}, t)] n_1(\mathbf{x})\, d\mathbf{x} \tag{2.5}$$

is the dynamic molecule number variable in the embryo and g is the relevant generalized coordinate.

A distribution function of an embryo with respect to the size is given as the average

$$f(g, t) =< n(g) >= \mathrm{Sp}\,[\,\delta(g - N_I)\rho(t)] \tag{2.6}$$

where $\rho(t)$ is a total nonequilibrium distribution function of the system depending on coordinates and momenta of all molecules constituting the system, N is the total number of vapor molecules in the system, and

$$\mathrm{Sp}[\ldots] = \sum_{N^{\mathrm{I}}} \sum_{N} \frac{1}{N^{\mathrm{I}}!(N - N^{\mathrm{I}})!} \int d\Gamma[\ldots]$$

It follows from definition (2.6) that

$$\frac{\partial f(g,t)}{\partial t} = < \dot{n}(g) > \tag{2.7}$$

as the distribution function $\rho(t)$ satisfies the Liouville equation.

The generalized coordinate g is not a dynamic variable, so it does not depend on time. Thus, from equation (2.4) we get

$$\dot{n}(g) = -\frac{\partial}{\partial g}\delta(g - N^{\mathrm{I}})\dot{N}^{\mathrm{I}} \tag{2.8}$$

where

$$N^{\mathrm{I}} = -\frac{d}{dt}(N - N^{\mathrm{I}}) = -\int d\mathbf{x}[\Theta(\mathbf{x},t)\dot{n}_1(\mathbf{x}) + \dot{\Theta}(\mathbf{x},t)n_1(\mathbf{x})] \tag{2.9}$$

The motion equation for the density $n_1(\mathbf{x})$ can be written in the form (2.5.4):

$$\dot{n}_1(\mathbf{x}) = -\operatorname{div}\mathbf{j}(\mathbf{x}) \tag{2.10}$$

where

$$\mathbf{j}(\mathbf{x}) = \sum_i \frac{\mathbf{p}_{1i}}{m_1}\delta(\mathbf{x} - \mathbf{r}_{1i}) \tag{2.11}$$

is the flow of the vapor particle number.

Substituting equation (2.10) into equation (2.9) and integrating by parts, we obtain

$$\dot{N}^{\mathrm{I}} = \int_{\Sigma} ds\,\mathbf{n}[\mathbf{j}(\mathbf{s}) - \mathbf{v}^S n_1(\mathbf{s})] \tag{2.12}$$

where we have used equation (2.6) and properties of the surface δ–function to reduce a volume integral to the integral over the surface Σ of the embryo. The velocity $\mathbf{v}^S$ differs insignificantly from the velocity of the embryo motion, which is henceforth set equal to zero. Thus

$$\dot{n}(g) = -\frac{\partial}{\partial g}n(g)\int_{\Sigma} ds\,\mathbf{n}\mathbf{j}(\mathbf{s}) \tag{2.13}$$

and equation (2.7) takes the form

$$\frac{\partial f(g,t)}{\partial t} = -\frac{\partial}{\partial g}I(g) \tag{2.14}$$

where

$$I(g) = \int_{\Sigma} ds\,\mathbf{n} < n(g)\mathbf{j}(\mathbf{s}) > \tag{2.15}$$

To derive a closed kinetic equation for $f(g,t)$ from equation (2.14) it is necessary to express the right–hand side of the equation (2.14) in terms of $f(g,t)$, and to find the required nonequilibrium distribution function $\rho(t)$.

We shall construct $\rho(t)$ following Zubarev's method, advanced by Bashkirov and Zubarev [41] for similar problems.

A quasiequilibrium distribution function for the system under consideration should include in the set $\{P_m\}$ (see Section 1.2) two time– and space–dependent parameters, as there are two nonequilibrium processes in the system, notably the phase transition (the embryo growth) and the vapor diffusion in the gas mixture. One of these parameters accouns for the nonequilibrium density of embryos in the g space, and another accounts for a nonequilibrium concentration of the vapor. Thus the quasiequilibrium distribution function [41] takes the form

$$\rho_q(t) = Q_q^{-1} \exp[-\beta\mathcal{H} + \int d\mathbf{x}\, \Theta \nu_1^{\mathrm{II}}(\mathbf{x},t) n_1(\mathbf{x}) + \int dg\, \varphi(g,t) n(g)] \quad (2.16)$$

where $\beta = 1/kT$, $\nu^{\mathrm{II}} = \beta\mu^{\mathrm{II}}$, and $\varphi(g,t)$ may be called the generalized chemical potential in space g; defined by the condition

$$< n(g) >=< n(g) >_q \quad (2.18)$$

This condition implies that the quasiequilibrium distribution $\rho_q(t)$ is such as to provide a coincidence of the average $< n(g) >_q$ with the exact value $< n(g) >$ obtained by means of the parameter $\varphi(g,t)$.

The nonequilibrium distribution function $\rho(t)$ accounting for dissipative processes in the system is to satisfy the Liouville equation

$$\frac{\partial \rho}{\partial t} + i\mathcal{L}\rho = -\varepsilon(\rho - \rho_q) \quad (2.19)$$

where $i\mathcal{L}$ is the operator of the Poisson brackets with the Hamiltonian (2.1), and the imaginary unit i is introduced to provide for the Hermitian character of the Liouville operator $\mathcal{L}$.

In accordance with the general method described in Chapter 1, the formal solution to equation (2.19) to a linear approximation in the deviation from ρ_q is

$$\begin{aligned} \rho(t) = \rho_q(t)\Big\{1 - \int d\mathbf{x}\, \Theta(\mathbf{x},t) \int_{-\infty}^{0} dt_1 \\ \times \Big[\int dg \frac{\partial \varphi(g,t+t_1)}{\partial g} n(g,t_1) - \nu^{\mathrm{II}}(\mathbf{x},t+t_1)\Big] \\ \times \operatorname{div}[\mathbf{j}(\mathbf{x},t_1) - < \mathbf{j}(\mathbf{x},t_1) >_q]\Big\} \end{aligned} \quad (2.20)$$

where time dependencies of $n(g,t_1)$ and $\mathbf{j}(\mathbf{x},t_1)$ correspond to a dynamic evolution in the form

$$n(g,t) = e^{i\mathcal{L}t}n(g)\,, \quad \mathbf{j}(\mathbf{x},t) = e^{i\mathcal{L}t}\mathbf{j}(\mathbf{x}) \tag{2.21}$$

4.3 Nucleation Kinetic Equation

The nonequilibrium distribution function (2.20) is used to calculate the average flow in the right–hand side of equation (2.14).

In the socalled local Markovian approximation, time and space variations of thermodynamic parameters are assumed to be much slower than the time and space decay of relevant correlation functions. So, while calculating the average flow we replace $\nu^{\mathrm{II}}(\mathbf{x}, t+t_1)$ from equation (2.20) by its value $\nu^{\mathrm{II}}(t)|_\Sigma$ on the embryo surface.

Then, we use the Gauss theorem to transform the integral taken over the gas–phase volume into the surface integral. Thus we get

$$\begin{aligned} \frac{\partial f(g,t)}{\partial t} = \frac{\partial}{\partial g}\Bigg\{ \left[\frac{\partial \varphi(g,t)}{\partial g} - \nu^{\mathrm{II}}|_\Sigma\right] \int_\Sigma d\mathbf{s} \int_\Sigma d\mathbf{s}' \int_{-\infty}^{0} dt_1\, \mathbf{n}\mathbf{n}' \\ \times < \delta(g-N^{\mathrm{I}})\mathbf{j}(\mathbf{s})\, [\mathbf{j}(\mathbf{s},t) - < \mathbf{j}(\mathbf{s},t_1) >_q] >_q \Bigg\} \end{aligned} \tag{3.1}$$

Let us consider an average of a dynamic function B,

$$< n(g)B >_q = \frac{\mathrm{Sp}\left[\delta(g-N_I)B\exp\left[-\beta\mathcal{H} + \int d\mathbf{x}\, \Theta\nu^{\mathrm{II}}(\mathbf{x})n(\mathbf{x}) + \varphi(N^{\mathrm{I}})\right]\right]}{\mathrm{Sp}\left[\exp\left[-\beta\mathcal{H} + \int d\mathbf{x}\, \Theta\nu^{\mathrm{II}}(\mathbf{x})n(\mathbf{x}) + \varphi(N^{\mathrm{I}})\right]\right]}$$

Multiplying and dividing this expression by

$$\sum_N \frac{1}{(N-g)!} \int d\Gamma \exp\left[-\beta\mathcal{H} + \int d\mathbf{x}\, \Theta\nu^{\mathrm{II}}(\mathbf{x})n(\mathbf{x})\right]$$

we find that

$$< n(g)B >_q = < B(g) >_l f(g,t) \tag{3.2}$$

where we have introduced an averaging $< \ldots >_l$ with the local–equilibribrium distribution

$$\rho_l(t|g) = Q_l^{-1}(g)\exp\left[-\beta\mathcal{H} + \int d\mathbf{x}\, \Theta\nu^{\mathrm{II}}(\mathbf{x})n(\mathbf{x})\right]$$

under the condition of a fixed number g of molecules in the embryo.

Using relation (3.2), equation (3.1) takes the form

$$\frac{\partial f(g,t)}{\partial t} = \frac{\partial}{\partial g}\left\{D(g)f(g,t)\left[\frac{1}{\beta}\frac{\partial\varphi(g,t)}{\partial g} - \mu^{\mathrm{II}}|_{\Sigma}\right]\right\} \tag{3.3}$$

where

$$D(g) = \beta\int_{\Sigma} d\mathbf{s}\int_{\Sigma} d\mathbf{s}'\int_{-\infty}^{0} dt\,\mathbf{nn} :< \mathbf{j}(\mathbf{s})\,[\mathbf{j}(\mathbf{s}',t) - < \mathbf{j}(\mathbf{s}',t) >_l] >_l \tag{3.4}$$

is the nucleation coefficient represented in the form of the time and surface integral of a density flow autocorrelation function.

Thus we have derived the Green–Kubo formula (3.4) for the nucleation coefficient. A further analysis of this formula is found in Section 4.4.

To close equation (3.3) it is necessary to express the parameter $\varphi(g,t)$ in terms of the distribution function $f(g,t)$ and conventional thermodynamical parameters determining a state of the system of the gas mixture and the embryo. The parameter $\varphi(g,t)$ is defined by equation (2.18). By means of writing its right–hand side in terms of the distribution function (2.16), and differentiating the logarithm of the resulting expression with respect to g, we find that

$$\frac{\partial \ln f(g,t)}{\partial g} = \frac{\partial\varphi(g,t)}{\partial g} + \frac{\partial \ln Q_l(g)}{\partial g}$$

On the other hand, we have

$$\frac{1}{\beta}\frac{\partial \ln Q_l(g)}{\partial g} = \frac{1}{\beta}\left[\frac{\partial \ln Q_l(g)}{\partial g}\right]_{g\to\infty} + \frac{1}{\beta}\frac{\partial \ln Q_l(g)}{\partial A(g)}\frac{\partial A(g)}{\partial g}$$

where the derivative with respect to g is presented as the sum of a volume part, calculated neglecting any influence of the embryo surface, and a surface part proportional to the change in $A(g)$. In thermodynamcs these derivatives of the free energy are the definitions of chemical potential, μ, and surface tension coefficient γ, respectively. So we have

$$\frac{1}{\beta}\frac{\partial\varphi}{\partial g} = \frac{1}{\beta}\frac{\partial \ln f(g,t)}{\partial g} + \mu_1^{\mathrm{I}} + \gamma\frac{\partial A(g)}{\partial g}$$

Thus, the nucleation kinetic equation takes the final form

$$\frac{\partial f(g,t)}{\partial t} = \frac{\partial}{\partial g}D(g)\left[\frac{\partial\Delta\Phi}{\partial g}f(g,t) + kT\frac{\partial f(g,t)}{\partial g}\right] \tag{3.5}$$

where $D(g)$ is defined by equation (3.4) and

$$\Delta\Phi = (\mu^{\mathrm{I}} - \mu^{\mathrm{II}}|_{\Sigma})g + \gamma A(g) \tag{3.6}$$

is the thermodynamic potential of nucleation. It differs from the classical expression, equation (1.2) in the value of the vapor chemical potential. Here, we have the value in the vicinity of the embryo surface, while in equation (1.2) it is the chemical potential for an undisturbed vapor at infinity. These two values are different if an embryo size is comparable with the mean free path in the gas mixture. For example, on the surface of a very large embryo the vapor chemical potential is equal to a chemical potential for the saturated vapor $\mu_1^*(R)$, where $R = (3gv_l/4\pi)^{1/3}$ is the embryo radius, v_l is the embryo volume for one molecule and there is a phase equilibrium on the phase boundary, i.e., $\partial\Delta\Phi/\partial g = 0$.

From the beginning the embryo is considered to be immobile, as it was previously understood in classical theory. Lothe and Pound [42] evaluated a correction for the classical value $\Delta\Phi$, taking into account translational and rotational movement of an embryo. This correction leads to the additional multiplier of 10^{17} to the classical value $\Delta\Phi$, which thus rejects classical nucleation theory. Thus, it is no wonder that the Lothe–Pound correction initiated a controversial discussion (see references on the subject in [43]) of the way of accounting for embryo movement. The author [44, 45] considers this movement as a Brownian thermal motion due to fluctuations of a momentum exchange with the surrounding gas mixture; thus the thermodynamical potential increase corresponding to this motion is irrelevant to the thermodynamics of a phase transition. Then, we the can disregard embryo thermal motion in nucleation theory.

4.4 Nucleation Coefficient

The Green–Kubo formula for the nucleation coefficient, equation (3.4), is remarkably compact as are all kinetic coefficients.

This formula is extremely general, in the sense that no specific approximation is made regarding the type of interactions. Precisely because of of this generality, explicit evaluation of the formulais very difficult: one is faced with the full N–body problem. Therefore, we shall henceforth confine our analysis to two limiting cases—very small and very large embryos in comparison with the mean free path in the gas mixture.

In the first of these two limiting cases (very small embryos), the intermolecular interaction in the gas phase may be neglected. Thus, the Liouville and evolution operators take the forms

$$i\mathcal{L} = \sum_{\alpha i} \frac{\mathbf{p}_{\alpha i}}{m_\alpha} \frac{\partial}{\partial \mathbf{r}_{\alpha i}} \qquad e^{i\mathcal{L}t} = 1 + i\mathcal{L} + \frac{1}{2!}(i\mathcal{L})^2 + \dots$$

and from equation (2.21) one obtains

$$\mathbf{j}(\mathbf{s}',t) = \sum_i \frac{\mathbf{p}_{1i}}{m_1}\delta(\mathbf{s}' - \mathbf{r}_{1i}(t)) = \sum_i \frac{\mathbf{p}_{1i}}{m_1}\delta(\mathbf{s}' - \mathbf{r}_{1i} - (\mathbf{p}_{1i}/m_1)_r t)$$

where $(\ldots)_r$ is the projection on the $\mathbf{r}_{1i}$ direction.

Substituting this expression into equation (3.4) and omitting subscripts α, i we get

$$D(g) = \frac{N\beta}{m^2}\int_\Sigma d\mathbf{s}\int_\Sigma d\mathbf{s}'\int_{-\infty}^{0} dt < p_n p_{n'}\delta(\mathbf{s}-\mathbf{r})\delta(\mathbf{s}'-\mathbf{r}-p_n\mathbf{n}t/m) >_l$$

Using the identity [14]

$$\int dx\,\delta(x-a)\delta(x-b) = \delta(a-b)$$

we have

$$D(g) = \frac{n\beta}{m^2}\int_\Sigma d\mathbf{s}\int_\Sigma d\mathbf{s}'\int_{-\infty}^{0} dt < p_n^2\delta(\mathbf{s}-\mathbf{s}'-p_n\mathbf{n}t/m) >_0 \qquad (4.1)$$

where the brackets $< \ldots >_0$ denote the mean value of function $\mathbf{p}$, calculated using the Maxwell distribution function, and $n = N/V$ is the mean–number density of vapor.

Because of the convexity of the embryo surface Σ, no two of its different points $\mathbf{s}$ and $\mathbf{s}'$ can be connected by a straight–line path $\mathbf{r}(t) = \mathbf{s}' - p_n\mathbf{n}t/m$ of a molecule of the vapor phase. Therefore the points $\mathbf{s}$ and $\mathbf{s}'$ in the integrand of equation (4.1) coincide.

Then equation (4.1) takes the form

$$\begin{aligned} D(g) &= \frac{n\beta}{m^2}\int_\Sigma d\mathbf{s}\int_\Sigma d\mathbf{s}'\int_{-\infty}^{0} dt < p_n^2\delta(\mathbf{s}-\mathbf{s}')\delta(p_n\mathbf{n}t/m) >_0 \\ &= \frac{n\beta A(g)}{m^2}\int_{-\infty}^{0} dt\,\delta(t) < p_n^2|m/p_n| >_0 = \frac{n\beta A(g)}{2m} < |p_n| >_0 \end{aligned}$$

where the identities

$$\delta(ax) = \delta(x)/|a| \quad \text{and} \quad \int_{-\infty}^{0} dt\,\delta(t) = \frac{1}{2}$$

were used [14].

Calculating the mean value of the momentum normal–component modulus, we finally obtain

$$D(g) = \frac{nA(g)}{(2\pi mkT)^{1/2}} \tag{4.2}$$

We have thus derived the classical expression of the nucleation coefficient in the free molecular regime from the general Green–Kubo formula (3.4).

Now, we discuss the second limiting case, a very large embryo with a radius R much greater then the mean free path in the gas mixture. There is a pure diffusion transfer of vapor to (from) a growing (decaying) embryo in this case. Then, the vapor concentration satisfies the Laplace equation for the stationary approximation

$$\nabla^2 \varrho^{\mathrm{II}}(\mathbf{x}) = 0 \tag{4.3}$$

with boundary conditions

$$\varrho^{\mathrm{II}}|_\Sigma = \varrho^*(R), \qquad \varrho^{\mathrm{II}}|_\infty = \rho|_\infty$$

where ϱ_∞ is the undisturbed vapor concentration and $\varrho^*(R)$ is the concentration of a relatively saturated vapor an embryo of radius R.

The spherically symmetrical solution for equation (4.3) satisfying these boundary conditions is

$$\varrho^{\mathrm{II}}(r) = \frac{R}{r}(\varrho^* - \varrho_\infty) + \varrho_\infty \tag{4.4}$$

To evaluate the coefficient $D(g)$ we multiply equation (3.4) by the vapor density $\varrho^{\mathrm{II}}|\Sigma$ on the embryo surface and convert the second surface integral into the volume using Gauss's theorem. Neglecting a surface integral taken over a closed surface surrounding the gas volume at infinity, we get

$$\varrho^{\mathrm{II}}|_\Sigma D(g) = \beta \int_\Sigma d\mathbf{s}\,\mathbf{n} \int d\mathbf{x}' \int_{-\infty}^{0} dt \frac{\partial}{\partial \mathbf{x}'} \varrho^{\mathrm{II}}(\mathbf{x}') < \mathbf{j}_1 \mathbf{s})[\mathbf{j}_1(\mathbf{x}',t) - < \mathbf{j}_1(\mathbf{x}',t) >] >_l$$

or, in the local approximation

$$\varrho^{\mathrm{II}}|_\Sigma D(g) = \int_\Sigma d\mathbf{s}\,\mathbf{n} \frac{\partial \varrho^{\mathrm{II}}}{\partial \mathbf{x}} D_1 - \int_\Sigma d\mathbf{s}\,\mathbf{n} \frac{\partial D_1}{\partial \mathbf{x}} \varrho^{\mathrm{II}}(\mathbf{x}) \tag{4.5}$$

where

$$D_1 = \frac{\beta}{3} \int d\mathbf{x}' \int_{-\infty}^{0} dt < \mathbf{j}_1 \mathbf{x})[\mathbf{j}_1(\mathbf{x}',t) - < \mathbf{j}_1(\mathbf{x}',t) >_l] >_l$$

is the diffusion coefficient of the vapor.

In the limiting case of very large R, the second term on the right–hand side of equation (4.5) should be omitted, since kinetic coefficients are assumed to be constant in the hydrodynamic approximation. Then, taking into account equations (4.3) and (4.4), we have

$$D(g) = 4\pi R D_1 \frac{\varrho_\infty - \varrho^*(R)}{\varrho^*(R)} \tag{4.6}$$

This expression is well known; it was first derived by Maxwell.

There is a phase equilibrium on the phase boundary, i.e., $\partial\Delta\Phi/\partial g = 0$, and equations (3.5) and (4.6) give the final form of the Becker equation for a large embryo:

$$\frac{\partial f(g,t)}{\partial t} = 4\pi k T D_1 \frac{\partial}{\partial g} R(g) \frac{\varrho_\infty - \varrho^*(R)}{\varrho^*(R)} \frac{\partial f(g,t)}{\partial g}$$

4.5 Entropy Production

A nonequilibrium entropy is defined as

$$\begin{aligned} S(t) &= k < \ln \rho_q(t) >= k \ln Q_q(t) + \frac{1}{T} < H > \\ &- \int d\mathbf{x}\, \Theta(\mathbf{x},t)\nu^{\mathrm{II}} < n(\mathbf{x}) > -k \int dg \varphi(g,t) f(g,t) \end{aligned} \tag{5.1}$$

By differentiating this expression, we get

$$\frac{dS}{dt} = -k \int d\mathbf{x}\, \Theta(\mathbf{x},t)\nu^{\mathrm{II}}(\mathbf{x},t) < \dot{n}(\mathbf{x}) > -k \int dg\, \varphi(g,t) \frac{\partial f(g,t)}{\partial t} \tag{5.2}$$

Using equations (2.10) and (2.14), we transform equation (5.2) to

$$\begin{aligned} \frac{dS}{dt} &= -k \int dg \left[\frac{\partial \varphi(g,t)}{\partial g} - \nu^{\mathrm{II}}|_\Sigma \right] I(g) \\ &-k \int d\mathbf{x}\, \Theta(\mathbf{x},t) \frac{\partial \nu^{\mathrm{II}}}{\partial \mathbf{x}} < \mathbf{j}(\mathbf{x}) > +k \int dg \frac{\partial}{\partial g} \left[\varphi(g,t) I(g) \right] \end{aligned} \tag{5.3}$$

We choose the Becker–Doering boundary condition $f(G,t) = 0$, where G is the maximum permissible size of the embryo (when an embryo reaches this size we withdraw it from the system). After differentiating expression (2.18) with respect to g, we have

$$\left.\frac{\partial f(g,t)}{\partial g}\right|_{g=G} = \left[f(g,t)\left(\frac{\partial \varphi}{\partial g} - \frac{\partial \ln Q_q}{\partial g}\right)\right]_{g=G} = 0$$

Hence it follows that

$$I(G) = -D(G)\left[\frac{\partial \Delta\Phi}{\partial g} f(g,t) + kT\frac{\partial f(g,t)}{\partial g}\right]_{g=G} = 0,$$

From equation (2.14) we find

$$\frac{\partial}{\partial t}\int_{g_0}^{G} dg\, f(g,t) = -I(G) + I(g_0) = 0 \tag{5.4}$$

where $I(g_0) = I(G) = 0$, and g_0 is a minimal embryo size, according to classical nucleation theory.

Thus, recalling the Becker–Doering boundary condition, we rewrite equation (5.3) in the form

$$\frac{dS}{dt} = \frac{1}{T}\int dg\, f(g,t)D(g)\left[\frac{\partial \Delta\Phi}{\partial g} + kT\frac{\partial \ln f}{\partial g}\right]^2 - k\int d\mathbf{x}\, \Theta(\mathbf{x},t)\frac{\partial \nu^{\mathrm{II}}}{\partial \mathbf{x}} < \mathbf{j}(\mathbf{x}) > \tag{5.5}$$

The first term on the right–hand side describes entropy production due only to the phase transition, and the second term describes entropy production from diffusion of vapor to the embryo. Both terms are positive definite and independent of the process direction (growth or evaporation of an embryo). The positive definition of the second term becomes evident when the diffusion flow is presented as

$$< \mathbf{j}(\mathbf{x}) >= -D_1\frac{\partial \nu^{\mathrm{II}}}{\partial \mathbf{x}}$$

If we depart from the Becker–Doering boundary condition, we obtain $I(g_0) = I(G) \neq 0$ in accordance with equation (5.4). Then additional terms will appear on the right–hand side of equation (5.5), these will be proportional to the difference of entropies introduced into the system by the flow $I(g_0)$ and withdrawn from the system by the flow $I(G)$.

4.6 Second Quantization Approach

Here we consider another approach to the nucleation theory applicable to the specific case of a rarefied gas–vapor mixture in which the free path length is much greater the critical embryo size.

In the free molecular regime, embryo growth does not cause any perturbations of vapor density. So there is no need to include motion of molecules of the gas or embryoin a dynamical description. Only molecular transfers from the vapor into the embryo and vice versa should be included. To do this, the second quantization method of quantum mechanics is preferred over the usual Hamiltonian classical mechanics. In the former method, transfer of a molecule from the vapor into the embryo is described simply as an annihilation of a molecule in the vapor and simultaneous creation of a molecule in the embryo, irrespective of the way the molecule moves.

The second quantization method is well developed now and widely used in quantum statistical mechanics, so here we will not go into details.

In the framework of this method, the state of a macroscopic system in the momentum representation is determined as a superposition of states, with fixed sets of numbers of particles possessing the momentum **p** (**p** takes all possible values). A secondary quantization operator a_p^+ (or a_p) creates (or annihilates) a particle with the momentum **p** changing the state of the system.

There are three subsystems in our case: embryo; vapor; and carrier gas, so we introduce three types of operators for creation a_p^+, b_p^+, and c_p^+, and annihilation a_p, b_p, and c_p of the particles of the embryo, vapor, and carrier gas, respectively. The commutation relations take the form

$$[a_p, a_{p'}^+]_\mp = [b_p, b_{p'}^+]_\mp = [c_p, c_{p'}^+]_\mp = \delta_{p',p} \tag{6.1}$$

and all remaining commutators (or anticommutators) vanish. The choice of the sign in these formulas is immaterial, since a nucleation process generally occurs at temperatures much higher than those of quantum degeneracy, i.e., classical Boltzmann statistics still holds.

The total Hamiltonian of the system can be written in the form

$$\mathcal{H} = H_e + H_t + V + \mathcal{T} \tag{6.2}$$

where

$$H_e = \sum_p \frac{p^2}{2m} a_p^+ a_p + \sum_{p_1 p_2 p_3 p_4} u(p_1 p_2 p_3 p_4) a_{p_1}^+ a_{p_2}^+ a_{p_3}^+ a_{p_4} \tag{6.3}$$

is the Hamiltonian of the embryo with the binary potential u of the intermolecular interaction (the embryo has volume v_g);

$$H_t = \sum_p \frac{p^2}{2m} b_p^+ b_p + \sum_p \frac{p^2}{2m_1} c_p^+ c_p \tag{6.4}$$

is the Hamiltonian of the thermostat, which is an ideal mixture of a vapor and carrier gas with molecular masses m and m_1, respectively; V is the

Hamiltonian of the elastic mechanical interaction between the embryo and the thermostat; and

$$\mathcal{T} = \sum_{pp'} (\mathcal{T}(g, p, p') a_p^+ b_{p'} + \mathcal{T}^*(g, p, p') b_{p'}^+ a_p) \tag{6.5}$$

is the tunneling Hamiltonian, which describes the transition processes of molecules from the vapor into the embryo and vice versa.

This form of the Hamiltonian for two systems separated by a potential barrier is well known in the theory of the Josephson effect; it was derived in the most general form by Prange [46]. Prange proved in the first approximation for the weak process, $\mathcal{T}$, the operators a and b describing the same particles on different sides of the potential barrier can be assumed to commute (anticommute) and the tunneling operator can be written in the form (6.5). A similar Hamiltonian is also used in the theory of chemical reactions. The explicit form of the matrix elements of $\mathcal{T}(g, p, p')$ in (6.5) depends on the actual form of the potential barrier and is specified by the choice of the embryo model. The argument g, the size (number of particles) of the embryo, reflects the dependence on g of the surface tension energy, i.e., the height and form of the barrier. The value of $\mathcal{T}$, which determines the intensity of the transition processes of molecules from the vapor into the embryo and vice versa, is assumed to be small and it will be the small parameter of the nucleation theory.

As in Section 4.2 we now introduce the density operator in the size space of the embryo

$$n(g) = \delta(g - N) \tag{6.6}$$

where $N = \sum_p a_p^+ a_p$ is the size operator (number of particles) of the embryo and g is a c number, an eigenvalue of the operator N. It is obvious that the eigenvalues g form a discrete set, but for the study of the behavior of a sufficiently smooth function $f(g, t)$, one can initially assume that g varies continuously.

The equation of motion for $n(g)$ in the linear approximation in $\partial/\partial g$ has the form

$$\dot{n}(g) = -\frac{\partial}{\partial g} \jmath(g) \tag{6.7}$$

where

$$\jmath(g) = n(g)\dot{N}, \quad \dot{N} = \frac{1}{i\hbar}[N, \mathcal{H}]_- = \frac{1}{i\hbar}[N, \mathcal{T}]_- \tag{6.8}$$

The exact equation of motion for $n(g)$ is a series in powers of $\partial/\partial g$ but the terms of higher order in $\partial/\partial g$ give contributions of higher order in $\mathcal{T}$, as was proved in the most general form in the investigation of Bashkirov and Zubarev [41]. Averaging the equation of motion (6.7) with a statistical operator satisfying the Louville equation with the Hamiltonian (6.1) gives the equation

$$\frac{\partial f(g,t)}{\partial t} = -\frac{\partial}{\partial g} < j(g) > \tag{6.9}$$

here

$$f(g,t) = < n(g) > \tag{6.10}$$

is the distribution function of the embryo with respect to the size (or, rather, with respect to the eigenvalues g of N) and the averaging $< \ldots >$ is performed over all dynamic variables of the system (the embryo and thermostat), including the size of the embryo N.

Our problem again reduces to calculating the mean flux on the right–hand side of equation (6.9), but for this purpose we must know the explicit form of the nonequilibrium statistical operator with which the averaging is performed.

4.7 Nonisothermal Nucleation Theory

Previously we did not allow for the latent heat of a phase transition due to which the embryo temperature can differ from the gas temperature. From physical considerations it is clear for a growing liquid embryo that a condensation precedes an evaporation, so the embryo temperature increases and as a result the rate of growing decreases. This effect is well known for systems where a rapid growth of embryos takes place, e.g., in combustion products effluxes from jet engines. However, it is only recently that publications on the topic have started appearing.

The system under consideration consists of the nonequilibrium subsystem, that is, the liquid embryo and the thermodynamic equilibrium gas mixture of the vapor and a neutral gas–carrier.

Unlike the isothermal nucleation theory discussed in the preceding sections, we now allow for the nonequilibrium size–dependent temperature of an embryo $T_e(g,t)$. Thus, an additional parameter of the contracted description is to be introduced. It is an embryo energy density in g–space:

$$h(g) = H_e n(g) \tag{7.1}$$

Its corresponding equation of motion can be written in the form

$$\dot{h}(g) = \dot{H}_e n(g) - \frac{\partial}{\partial g}(H_e j(g)) \tag{7.2}$$

where

$$\dot{H}_e = \frac{1}{ih}[H_e, \mathcal{T} + V] \tag{7.3}$$

The nonequilibrium statistical operator that is the solution of the Liouville equation with the Hamiltonian (6.2) in a linear flow approximation (i.e., $\mathcal{T}_{pq}$) becomes

$$\begin{aligned}\rho(t) = \rho_l(t) + \int_0^\beta d\lambda \int_{-\infty}^0 dt \Big[\int dg\, \alpha(g,t) \dot{H}_e(t_1,\lambda) n(g) \\ - \dot{H}_T(t_1,\lambda) + \int dg \frac{\partial \alpha(g,t+t_1)}{\partial g} H_e(t_1,\lambda) j(g,t_1,\lambda) \\ - \int dg \frac{\partial \varphi(g,t+t_1)}{\partial g} j(g,t_1,\lambda) \Big] \rho_l(t)\end{aligned} \tag{7.4}$$

where

$$\rho_l(t) = Q_l^{-1} \exp\{-\beta[H_T + V + \mathcal{T} + \int dg\, (\alpha(g,t)H_e - \varphi(g,t))(g)]\} \tag{7.5}$$

The parameter $\varphi(g,t)$ is here the same as in Section 4.2 ; $\beta = 1/kT$, where T is the thermostat temperature, and $\alpha(g,t) = T/T_e(g,t)$.

Equation (6.9) for $f(g,t)$ becomes

$$\begin{aligned}\frac{\partial f(g,t)}{\partial t} = -\frac{\partial}{\partial g}\{f(g,t)[(\alpha(g,t)-1)L_{12} \\ + \frac{\partial \alpha(g,t)}{\partial g} L_{1H1} - \frac{\partial}{\partial g}[\Delta\Phi(g,\alpha,\beta) + kT \ln f(g,t)] L_{11}\}\end{aligned} \tag{7.6}$$

where

$$\Delta\Phi(g,\alpha,\beta) = \Delta\Phi_{is} + (\alpha(g,t)-1)F_c(\beta,g) \tag{7.7}$$

Here $\Delta\Phi_{is}$ is the isothermal free energy of the embryo formation and $F_c(\beta,g)$ is the equilibrium free energy of the nucleus; the parameter $\varphi(g,t)$ has been excluded here as in Section 4.3. The kinetic coefficients are equal

$$L_{11} = \int_0^\beta d\lambda \int_{-\infty}^0 dt < \dot{N}\dot{N}(t_1,\lambda) >_q \tag{7.8}$$

$$L_{12} = \int_0^\beta d\lambda \int_{-\infty}^0 dt < \dot{N}\dot{H}_e(t_1,\lambda) >_q \tag{7.9}$$

$$L_{1H1} = \int_0^\beta d\lambda \int_{-\infty}^0 dt < \dot{N}H_e\dot{N}(t_1,\lambda) >_q \tag{7.10}$$

where $< \ldots >_q$ is the average over the statistical operator

$$\rho_q = Q_q^{-1} \exp\{-\beta[H_T + H_e + \mathcal{T} + V - \varphi(g,t)]\} \tag{7.11}$$

The equation presenting the change of $\alpha(g,t)$ with time was derived by averaging equation (7.3) as

$$
\begin{aligned}
&f(g,t)\frac{\partial\alpha(g,t)}{\partial t}\int_0^\beta d\lambda\,[< H_e H_e(\lambda) >_q - < H_e >_q< H_e(\lambda) >_q] \\
&= -\left[(\alpha(g,t)-1)L_{22} + \frac{\partial\alpha(g,t)}{\partial g}L_{2H1} - \beta\frac{\partial\varphi(g,t)}{\partial g}\right] f(g,t) \\
&+ < H_e >_q \frac{\partial f(g,t)}{\partial t} + \frac{\partial}{\partial g}\Big\{f(g,t)\Big[(\alpha(g,t)-1)L_{H12} \\
&+ \frac{\partial\alpha(g,t)}{\partial g}L_{H1H1} - \frac{\partial\varphi(g,t)}{\partial g}L_{H11}\Big]\Big\}
\end{aligned}
\tag{7.12}
$$

where

$$L_{22} = \int_0^\beta d\lambda \int_{-\infty}^0 dt < \dot{H}_e\dot{H}_e(t_1,\lambda) >_q \tag{7.13}$$

$$L_{H22} = \int_0^\beta d\lambda \int_{-\infty}^0 dt < H_e\dot{N}\dot{N}(t_1,\lambda) >_q \tag{7.14}$$

The form of the other kinetic coefficients is easy to restore by comparision with the preceding ones. In these expressions, $\dot{N}$ and $\dot{H}_e$ are the rates of change of the molecular number and the energy of the embryo, respectively. The timedependence of $\dot{N}$ and $\dot{H}_e$ on t_1 and λ is defined as

$$
\begin{aligned}
\dot{N}(t_1,\lambda) &= \exp\{\lambda(H_T+\alpha H_e+\mathcal{T}+V)\}\exp\{i\mathcal{H}t_1/\hbar\}\dot{N} \\
&\times\exp\{-\lambda(H_T+\alpha H_e+\mathcal{T}+V)\}\exp\{-i\mathcal{H}t_1/\hbar\}
\end{aligned}
\tag{7.15}
$$

4.8 The Model Example

For the practical application of equations (7.6) and (7.12), representing the growth of the nucleus and its heat exchange with the thermostat, it is necessary to know the explicit form of the kinetic coefficients included. In particular, they may be calculated precisely from the simplest model of a nucleus where the pair intermolecular interaction is substituted for the average field.

Although this model is too rough to represent liquid drop thermodynamics, which are strongly depended on local deviations of the intermolecular interaction field from average, it is useful for considering the nucleation kinetics.

In fact, if the liquid is considered as an ideal gas in the potential field U formed by the averaged intermolecular interaction, then the relation c^*/c_l between the densities of the saturated vapor and the liquid in the equilibrium state of the gas–vapor system must obey the equation $c^* = c_l \exp\{-\beta U\}$, whence

$$U = -kT \ln c^*/c_l \tag{8.1}$$

Thus, within the scope of the given model, the latent heat U of the phase transformation may be determined, based on the known values c^* and c_l with the fixed value β. In particular, for water at $273K$, we obtain a latent heat equal to 350 cal/g. This value is $\sim 60\%$ of the experimental value for the macroscopic liquid.

It is well known that the latent heat for small systems is less than that for macroscopic ones. Thus, one should expect that, for the nuclei, the practical correspondence to the experimental data will be better.

To obtain the exact solution of the model example, we consider a system consisting only of a vapor and a droplet (without carrier gas) and with the condensation coefficient assumed to be equal to unity. Then

$$H_e = \sum_p \left(\frac{p^2}{2m} - U\right) a_p^+ a_p = \sum_p \varepsilon_p a_p^+ a_p \tag{8.2}$$

$$H_T = \sum_q \frac{q^2}{2m} b_q^+ b_q = \sum_q E_q b_q^+ b_q \tag{8.3}$$

As we allow only for terms at most of second order in $\mathcal{T}$, the evolution of the operators N and H_e may be considered free, i.e., in neglecting the interaction Hamiltonian in the Heisenberg representation (7.15), for the operators $\dot{N}$ and $\dot{H}_e$.

Considering the temperature difference $(T - T_c)$ to be small, we pass from averaging with the quasiequilibrium statistical operator ρ_q to averaging with the operator

$$\rho_0 = Q_0^{-1} \exp\{-\beta(H_T + H_e)\} \tag{8.4}$$

It is not difficult to calculate the kinetic coefficients using Wick theorem by substituting of ρ_0 for ρ_q and using the model Hamiltonians, equations (8.2) and (8.3). Because the temperature range in question is much higher than the quantum degeneration temperature, we may confine consideration to only terms not higher than first order terms in the occupation numbers

$$f_p = < a_p^+ a_p > \ll 1\,, \quad f_q = < b_q^+ b_q > \ll 1 \tag{8.5}$$

Then,

$$L_{11} = \beta\pi/\hbar \sum_{pq} \mathcal{T}_{pq}^2 (f_p + f_q)\delta(\varepsilon_p - E_q) \tag{8.6}$$

$$L_{12} = \beta\pi/\hbar \sum_{pq} \varepsilon_p \mathcal{T}_{pq}^2 (f_p + f_q)\delta(\varepsilon_p - E_q) \tag{8.7}$$

$$L_{12} = \beta\pi/\hbar \sum_{pq} \varepsilon_p \mathcal{T}_{pq}^2 f_p \delta(\varepsilon_p - E_q) \tag{8.8}$$

where

$$f_p = \exp\{-\beta(\varepsilon_p - \mu)\}, \quad f_q = \exp\{-\beta(E_q - \mu)\} \tag{8.9}$$

The chemical potential μ is defined from the condition

$$\sum_p f_p = g \tag{8.10}$$

Another analogous condition, $\sum f_q = N - g$, will then be fulfilled automatically as a consequence of the definition of the potential U.

Let us now determine the matrix elements $\mathcal{T}_{pq}$ within the scope of our model. The value $\frac{2\pi}{\hbar}\mathcal{T}_{pq}^2\delta(\varepsilon_p - E_q)$ in equations (8.6) through (8.8) is the probability for transition $p \rightleftharpoons q$ per unit time. Then

$$W = \frac{2\pi}{\hbar} \sum_p \mathcal{T}_{pq}^2 \delta(\varepsilon - E_q) \tag{8.11}$$

is the probability that a vapor molecule with momentum q will condense in the unit time. In all the equations hereafter, we switch from summing over quantum states to integrating with respect to momentum:

$$\sum_q \to \frac{4\pi\Omega}{(2\pi\hbar)^3} \int_0^\infty dq\, q^2 , \quad \sum_p \to \frac{4\pi V_g}{(2\pi\hbar)^3} \int_0^\infty dp\, p^2 \tag{8.12}$$

Then

$$W = \frac{2\pi}{\hbar} \frac{V_g 4\pi m}{(2\pi\hbar)^3} \sqrt{q^2 + 2mU}\, \mathcal{T}^2_{\sqrt{q^2+2mU},p} \tag{8.13}$$

Alternatively, in the free molecular regime, the probability that a molecule with momentum q in volume Ω will collide with an embryo of surface area A equals

$$W = qA/(4m\Omega) \tag{8.14}$$

By combining equations (8.13) and (8.14) we get

$$\mathcal{T}^2_{\sqrt{q^2+2mU},p} = \frac{q}{\sqrt{q^2 + 2mU}} \frac{A(2\pi\hbar)^3\hbar}{32\pi^2 m^2 \Omega V_g} \tag{8.15}$$

It is now possible to calculate all the kinetic coefficients. We illustrate the method of calculation using L_{11} as an example:

$$\begin{aligned} L_{11} &= \frac{32\pi^3\Omega V_g m}{(2\pi\hbar)^6\hbar}\int_0^\infty dp\,p^2\int_0^\infty dq\,q^2T_{pq}^2\delta(p-\sqrt{q^2+2mU})(f_p+f_q)/2p \\ &= \frac{2\pi mA\exp\{\beta\mu\}}{(2\pi\hbar)^3\beta} \end{aligned} \tag{8.16}$$

Here, the property of the δ function [14]

$$\delta(x^2-a^2)=\frac{\delta(x-a)+\delta(x+a)}{2|a|} \tag{8.17}$$

has been used.

By substituting into equation (8.16) the expression for the chemical potential of an ideal gas at pressure p, we get

$$L_{11}=nA/\sqrt{2\pi mkT}=D \tag{8.18}$$

where D is the classical expression (4.2) for the nucleation coefficient.

The following expressions are obtained in the same way:

$$\begin{aligned} &L_{12}=2kTD\,,\quad L_{H11}=L_{1H1}=kTD\,,\quad L_{1H1H}=O(f_p) \\ &L_{22}=6(kT)^2D\,,\quad L_{2H1}=L_{H12}=3(kT)^2D \end{aligned} \tag{8.19}$$

By substituting the expressions for the kinetic coefficients, the equations (8.18) and (8.19) into equations (7.6) and (7.12), we obtain the set of equations representing the nonisothermal nucleation process

$$\begin{aligned} \frac{\partial f(g,t)}{\partial t} = \frac{\partial}{\partial g}\Bigg\{ kTDf(g,t)\Bigg[\frac{\partial}{\partial g}\left(\frac{\Delta\Phi(g,\alpha,\beta)}{kT}+\ln f(g,t)\right) \\ -\frac{\partial\alpha(g,t)}{\partial g}-2\left(\alpha(g,t)-1\right)\Bigg]\Bigg\} \end{aligned} \tag{8.20}$$

$$f(g,t)\frac{\partial\alpha(g,t)}{\partial t}CkT = (kT)^2 f(g,t)D\Big[2\frac{\partial}{\partial g}\Big(\frac{\Delta\Phi(\alpha,g,t)}{kT} + \ln f(g,t)\Big) - 3\frac{\partial\alpha(g,t)}{\partial g} - 6\,(\alpha(g,t)-1)\Big] + \left(\frac{3}{2}kT - U\right) g\frac{\partial f(g,t)}{\partial t} - \frac{\partial}{\partial g}\Big\{(kT)^2 f(g,t)D \times\Big[\frac{\partial}{\partial g}\left(\frac{\Delta\Phi(g,\alpha,\beta)}{kT} + \ln f(g,t)\right) - 3\,(\alpha(g,t)-1)\Big]\Big\} \tag{8.21}$$

Here, we have substituted $< H_e^2 > - < H_e >^2 = (kT)^2 C$, where C is the nucleus heat capacity.

In conclusion, we note that the evolution of $f(g,t)$ and $\alpha(g,t)$ represented by these two equations is defined by the following measured values: T, c^*, c_l, C, the surface tension γ, and actual vapor pressure in the system, which enters into the expression for $\Delta\Phi$.

Thus, a simultaneous solution to this system of two equations gives the possibility of determining the size distribution function and the temperature of the embryos. It is quite a complicated problem even for this model, so we restrict ourselves to qualitative estimation.

If we assume without noting equation (8.21), that the temperature of the growing embryo exceeds the temperature of the vapor by only one degree, then the respective changes of $\Delta\Phi(g,\alpha)$ and the term with $(\alpha-1)$ in equation (8.20) result in a tenfold decrease in the critical embryo formation rate.

Sharaf and Dobbins [43] measured the nucleation rate as the pressure decreased abruptly (the ratio $\dot{p}/p$ was varied from 1 to $5\,\mathrm{s}^{-1}$). The results were compared with the different theoretical predictions. It was found that the classical Becker–Doering theory gave the best fit even though it overestimated the rate by one or two orders. According to the above estimates this discrepancy can be compensated for using the assumption that the temperature of an embryo was higher than the gas temperature by one or two degrees.

Chapter 5

Brownian Motion: Momentum Flow Fluctuations on Interphase Boundary

5.1 Introduction

In Chapter 2 we studied nonequilibrium statistical thermodynamics for a two–phase system with a surface phase. We considered the linear transport equations, the boundary conditions for the thermodynamic parameters and flows, and the general form of the corresponding kinetic coefficients. In Chapter 4 nucleation was studied as an example of fluctuation phenomena on an interphase boundary. Here we shall consider another characteristic interphase fluctuation phenomenon, namely, the Brownian motion of a large spherical particle in an incompressible viscous liquid.

Brownian motion was first observed in the early nineteenth century. Development of its theory favored a molecular–kinetic views of the process—that it had a heat nature and a substance structure. The molecular–kinetic theory of this phenomenon was developed in 1905 by Einstein. In his theory, the observed random motion of a suspended particle was explained by stochastic collisions between the particle and molecules of the liquid, and the damping of this motion was determined by the Stokes drag coefficient $\zeta_S = 6\pi r_0 \eta$, where r_0 is the radius of the Brownian particle, and η is the coefficient of shear viscosity of the liquid.

The present state of nonequilibrium thermodynamics makes it possible to refine this model. In accordance with the fluctuation–dissipation theorem, the Brownian motion of a large particle in a viscous incompressible liquid must be due to the fluctuation of the tensor of the viscous shear stresses, since the dissipative characteristic of this effect, the coefficient ζ_S, is expressed in terms of the viscosity η. This means that the random force acting on the Brownian particle arises as a result of the interaction of the particle surface with a kind of collective motion in the liquid corresponding to the fluctuations of the shear stresses. Thus, the tangential components of the forces of the interaction with individual molecules are correlated, which ensures that together they produce the effect. Essentially, such an approach was the basis of the phenomenological theory of Fox and Uhlenbeck [47], in which the random force acting on the particle was expressed in terms of the fluctuation of the tensor of the viscous stresses of an incompressible liquid, and the autocorrelation function of the random force was expressed in terms of the coefficient ζ_S.

The statistical theory of Brownian motion originates from Bogoliubov's and Krylov's work [48] on the theory of Brownian motion in a system of oscillators, and the work of Kirkwood [49], who derived from first principles a Fokker–Planck equation for the distribution function of a heavy Brownian particle in a liquid and an explicit expression of the Green–Kubo formula type for the drag coefficient

$$\zeta_k = \frac{1}{kT}\int_{-\infty}^{0} dt\ \langle \mathbf{F}\mathbf{F}(t)\rangle_0 \tag{1.1}$$

where k is Boltzmann's constant, T is the temperature of the liquid, and $\langle\ldots\rangle_0$ denotes averaging over the phase space of the liquid with the equilibrium Gibbs distribution function. The force $\mathbf{F}$ is the gradient of the potential $\sum_i U_i$ of the interaction of the Brownian particle with the molecules of the liquid,

$$\mathbf{F} = -\sum_i \frac{\partial U_i}{\partial \mathbf{R}} \tag{1.2}$$

where $\mathbf{R}$ is the coordinate of the Brownian particle.

In nonequilibrium statistical mechanics, many new methods have now been developed for obtaining an equation of the Fokker–Planck type as well as an expression for the drag of the type of (1.1) in more general cases and more rigorously (see, for example, [50, 51, 52] and their bibliographies). However, as usual, as in the Kirkwood paper [49], it is assumed that the interaction of the Brownian particle with the molecules of the liquid is described by a central potential U_i. Such a model does not enable one to account for the tangential component of the random force acting on the surface of the Brownian particle. It is not surprising that the expressions of type (1.1) obtained in such theories for the drag do not have anything in common with the Stokes coefficient ζ_S.

This is confirmed in particularly by the calculation using equation (1.1) for a heavy particle in a gas (see Section 5.9).

Thus, the problem of constructing a statistical theory for the Brownian motion of a large (macroscopic) particle remained long unsolved. The first paper on this subject [53] was based on strong model assumptions about the nature of the interaction between liquid and solid surfaces. This paper derived the Fokker–Planck equation and the expression for the drag coefficient resulting in the Stokes formula.

Here, on the basis of the general approach to the description of processes at the boundary between two phases developed in the Chapter 2, we obtain the same results without using additional assumptions about the interactions of the liquid with the surface of the Brownian particle. We will also discuss the Brownian motion in a hydrodynamically nonequilibrium fluid.

It is clear that under the influence of the fluctuation of the shear stresses, a large particle must execute not only translational but also rotational Brownian motion. However, in the present chapter, which is devoted to the development of the new formalism, we shall consider only translational motion.

5.2 Microscopic Description of the System of Brownian Particle and Fluid

The Hamiltonian of the system, which consists of a solid macroscopic particle and a K–component fluid, has the form

$$\begin{aligned}\mathcal{H} =& \sum_{\alpha}^{K}\sum_{i}\left[\frac{p_{\alpha i}^2}{2m_\alpha}+\frac{1}{2}\sum_{\beta,j}u^{\alpha\beta}(\mathbf{r}_{\alpha i}-\mathbf{r}_{\beta j})\right]\\ &+\sum_{k}\left[\frac{p_k^2}{2m_0}+\frac{1}{2}\sum_{l}u^0(\mathbf{r}_k-\mathbf{r}_l)\right]+\sum_{\alpha,i,k}u^{0\alpha}(\mathbf{r}_{\alpha i}-\mathbf{r}_k)\end{aligned} \tag{2.1}$$

where the subscripts α, β, i, and j denote the coordinates and momenta of the molecules of the liquid with mass m_α, m_β; the subscripts k, l denote the coordinates and momenta of the molecules of the solid particle with mass m_0; $u^{\alpha\beta}$ is the potential of the interaction between the molecules of the fluid; u^0 is the potential of the interaction between the molecules of the solid particle; and $u^{0\alpha}$ is the potential of the interaction between the molecules of the fluid and the molecules of the particle.

The fluid is in a hydrodynamically nonequilibrium state. The movement of the particle generates an additional disturbance of the fluid state. As we consider the fluid and the solid particle as two bulk phases I and II, respectively we can write the quasiequilibrium distribution function of the system in the form of equation (2.3.5). The physical sense of the surface phase **S** in this case will be discussed later.

Since we are ultimately interested in a description of the motion of the particle as a whole, we introduce collective variables for the momenta and coordinate of the center of mass:

$$\mathbf{P} = \sum_k \mathbf{p}_k \,, \quad \mathbf{R} = \frac{1}{N_0} \sum_k \mathbf{r}_k \tag{2.2}$$

where N_0 is the number of molecules in the particle. We introduce the dynamic density

$$n(\mathbf{P}, \mathbf{R}) = \delta \left(\mathbf{P} - \sum_k \mathbf{p}_k \right) \delta \left(N_0 \mathbf{R} - \sum_k \mathbf{r}_k \right) \tag{2.3}$$

The mean of this quantity over the complete phase space of the system Γ with the total nonequilibrium distribution function $\rho(\mathbf{p}_i, \mathbf{r}_i, \mathbf{p}_k, r_k, t)$ is the distribution function of the phase coordinates of the center of mass of the Brownian particle:

$$f_B(\mathbf{P}, \mathbf{R}, t) = \mathrm{Sp}\left[n(\mathbf{P}, \mathbf{R}) \rho(\mathbf{p}_i, \mathbf{r}_i, \mathbf{p}_k, r_k, t) \right] \tag{2.4}$$

which we shall for brevity refer to as the distribution function of the Brownian particle.

We introduce the projection operator

$$\mathcal{P} A = \rho_q^{\mathrm{con}}(t) \mathrm{Sp}\left[n(\mathbf{P}, \mathbf{R}) A \right] \tag{2.5}$$

where A is an arbitrary function of the dynamic variables, and $\rho_q^{\mathrm{con}}(t)$ is the conditional quasiequilibrium distribution function of the system. This function differs from ρ_q, equation (2.3.5), by the condition that the coordinate and momentum of the center of mass of the Brownian particle are equal to the fixed **R** and **P**, and also by the fact that it is normalized with the density $n(\mathbf{P}, \mathbf{R})$, i.e.,

$$\mathrm{Sp}\left[n(\mathbf{P}, \mathbf{R}) \rho_q^{\mathrm{con}}(t) \right] = 1 \tag{2.6}$$

In particular, applying the operator $\mathcal{P}$ to $\rho(t)$, we obtain

$$\mathcal{P} \rho(t) = \rho_q^{\mathrm{con}}(t) f_B(\mathbf{P}, \mathbf{R}, t) \tag{2.7}$$

It is easy to show that the operator defined in this manner has the fundamental property of a projection operator: $\mathcal{P}^2 = \mathcal{P}$, $\mathcal{P}(1 - \mathcal{P}) = 0$.

5.3 Derivation of the Kinetic Equation

Using the projection operator, we can formulate a boundary condition for the Liouville equation (1.3.4) at $t = -\infty$:

$$\rho(t)|_{t\to-\infty} = [\mathcal{P}\rho(t)]_{t\to-\infty} \equiv [f_B(\mathbf{P},\mathbf{R},t)\rho_q^{\text{con}}(t)]_{t\to-\infty} \tag{3.1}$$

which assumes that at the initial time there were no correlations between the collective variables $\mathbf{P}$ and $\mathbf{R}$ and the remaining variables of the system that were fulfilled by the quasiequilibrium distribution $\rho_q^{\text{con}}(t)$. To take into account this boundary condition, equation (1.3.3) is written as

$$\frac{\partial\rho}{\partial t} + i\mathcal{L}\rho = -\varepsilon(\rho - \mathcal{P}\rho) \tag{3.2}$$

We apply the operator $\mathcal{P}$ to this equation:

$$\mathcal{P}\frac{\partial\rho}{\partial t} + i\mathcal{P}\mathcal{L}\rho = 0 \tag{3.3}$$

and consider the second term

$$\begin{aligned} i\mathcal{P}\mathcal{L}\rho &= \rho_q^{\text{con}}\text{Sp}\left[n(\mathbf{P},\mathbf{R})\left[\sum_{\alpha i}\left(\dot{\mathbf{r}}_{\alpha i}\frac{\partial}{\partial\mathbf{r}_{\alpha i}} + \dot{\mathbf{p}}_{\alpha i}\frac{\partial}{\partial\mathbf{p}_{\alpha i}}\right)\right.\right. \\ &\left.\left. + \sum_{\alpha k}\left(\dot{\mathbf{r}}_{\alpha k}\frac{\partial}{\partial\mathbf{r}_{\alpha k}} + \dot{\mathbf{p}}_{\alpha k}\frac{\partial}{\partial\mathbf{p}_{\alpha k}}\right)\right]\rho(t)\right] \\ &= \rho_q^{\text{con}}\text{Sp}\left[\rho(t)\left(\frac{\mathbf{P}}{M}\frac{\partial}{\partial\mathbf{R}} + \dot{\mathbf{P}}\frac{\partial}{\partial\mathbf{P}}\right)n(\mathbf{P},\mathbf{R})\right] \\ &= \rho_q^{\text{con}}\left[\frac{\mathbf{P}}{M}\frac{\partial f_b}{\partial\mathbf{R}} + \frac{\partial}{\partial\mathbf{P}} < \dot{\mathbf{P}} >_q^{\text{con}} f_B + \frac{\partial}{\partial\mathbf{P}}\text{Sp}\left[\dot{\mathbf{P}}n(\mathbf{P},\mathbf{R})\varphi(t)\right]\right] \end{aligned} \tag{3.4}$$

where in the last step we have used the identity $\rho = \mathcal{P}\rho + (1-\mathcal{P})\rho$ and introduced the notations for the correlation function $\varphi(t) = (1-\mathcal{P})\rho(t)$, for the averaging $< \ldots >_q^{\text{con}} = \text{Sp}\left[n(\mathbf{P},\mathbf{R})\ldots\rho_q^{\text{con}}\right]$, and for the mass M of the Brownian particle, $M = N_0 m_0$.

Substituting (3.4) in (3.3), we obtain

$$\frac{\partial f_B}{\partial t} + \frac{\mathbf{P}}{M}\frac{\partial f_b}{\partial\mathbf{R}} + \frac{\partial}{\partial\mathbf{P}}(< \dot{\mathbf{P}} >_q^{\text{con}} f_b) = -\frac{\partial}{\partial\mathbf{P}}\text{Sp}\left[\mathbf{P}n(\mathbf{P},\mathbf{R})\varphi(t)\right] \tag{3.5}$$

In accordance with Zwanzig's general projection operator method [54], we get the equation for $\varphi(t)$ by applying to (3.2) the projection operator $(1-\mathcal{P})$:

$$\frac{\partial\varphi}{\partial t} + \varepsilon\varphi + (1-\mathcal{P})i\mathcal{L}\varphi = -(1-\mathcal{P})i\mathcal{L}\mathcal{P}\rho - \frac{\partial\ln\rho_q^{\text{con}}}{\partial t}\mathcal{P}\rho \tag{3.6}$$

The last term on the right–hand side has arisen from the time dependence of the projection operator $\mathcal{P}$. The formal solution to this equation has the form

$$\varphi(t) = -\int_{-\infty}^{t} dt'\, e^{\varepsilon(t'-t)} U(t,t') \left\{ \left[\frac{\partial \ln \rho_q^{\rm con}}{\partial t'} + i(1-\mathcal{P})\mathcal{L} \right] \mathcal{P}\rho \right\}_{t'} \tag{3.7}$$

where $U(t,t')$ is the evolution operator determined by the equation

$$\frac{\partial U(t,t')}{\partial t} = -i(1-\mathcal{P})\mathcal{L}U(t,t') \tag{3.8}$$

with boundary condition $U(t,t) = 1$, the subscript t' of the curly brackets in (3.7) corresponds to the time argument of the distribution functions in the brackets.

Replacing the variable of integration t' in the integral (3.7) by $t_1 = t' - t$, we transform it to

$$\begin{aligned} \varphi(t) &= -\int_{-\infty}^{0} dt_1\, e^{\varepsilon t_1} U(t, t+t_1) \Big\{ \left(\frac{\partial \rho_q^{\rm con}}{\partial t_1} + i\mathcal{L}\rho_q^{\rm con} \right) f_B \\ &\quad + i(1-\mathcal{P})\rho_q^{\rm con}\mathcal{L} f_B \Big\}_{t+t_1} \end{aligned} \tag{3.9}$$

where we have used the following property of the differential operator $\mathcal{L}$:

$$i\mathcal{L}\mathcal{P}\rho = i f_B \mathcal{L}\rho_q^{\rm con} + i\rho_q^{\rm con}\mathcal{L} f_B \tag{3.10}$$

To calculate the action of the Liouville operation $\mathcal{L}$ on the function $f_B(\mathbf{P}, \mathbf{R}, t)$, it is necessary to express $\mathcal{L}$ and f_B in terms of identical variables. The simplest way of doing this is to express f_B as a function of the original variables $\mathbf{p}_k$ and $\mathbf{r}_k$. Then

$$\begin{aligned} i\mathcal{L} f_B &= \sum_k \left(\dot{\mathbf{r}}_k \frac{\partial}{\partial \mathbf{r}_k} + \dot{\mathbf{p}}_k \frac{\partial}{\partial \mathbf{p}_k} \right) f_B \left(\sum_k \mathbf{p}_k, \frac{1}{N} \sum_k \mathbf{r}_k, t \right) \\ &= \left(\dot{\mathbf{R}} \frac{\partial}{\partial \mathbf{R}} + \dot{\mathbf{P}} \frac{\partial}{\partial \mathbf{P}} \right) f_B(\mathbf{P}, \mathbf{R}, t) \end{aligned} \tag{3.11}$$

where we have again introduced the notation $\mathbf{R}$ and $\mathbf{P}$ for the center–of–mass phase variables.

The same result is obtained by a different approach based on expression $\mathcal{H}$ and $\mathcal{L}$ in terms of the individual variables and collective variables R and P (see, for example, Chapter 6 of the monograph [55]).

Thus,

$$i(1-\mathcal{P})\rho_q^{\rm con}\mathcal{L} f_b = \rho_q^{\rm con}(\dot{\mathbf{P}} - \langle \dot{\mathbf{P}} \rangle_q^{\rm con}) \frac{\partial f_b}{\partial \mathbf{P}} \tag{3.12}$$

Substituting this expression into equation (3.9) and the resulted formula obtained for $\varphi(t)$ into (3.5), we arrive at the exact non–Markovian kinetic equation for the distribution function of the Brownian particle:

$$\frac{\partial f_B}{\partial t} + \frac{\mathbf{P}}{M}\frac{\partial f_B}{\partial \mathbf{R}} + \frac{\partial}{\partial \mathbf{P}}(< \dot{\mathbf{P}} >_q^{\text{con}} f_B) = \frac{\partial}{\partial \mathbf{P}} \text{Sp} \left[n(\mathbf{P}, \mathbf{R}) \int\limits_{-\infty}^{0} dt_1 \, e^{\varepsilon t_1} \right.$$
$$\left. \times \dot{\mathbf{P}} U(t, t+t_1) \left\{ \left[\frac{\partial \rho_q^{\text{con}}}{\partial t_1} + i\mathcal{L}\rho_q^{\text{con}} \right] f_B + (\dot{\mathbf{P}} - < \dot{\mathbf{P}} >_q^{\text{con}}) \frac{\partial f_B}{\partial \mathbf{P}} \rho_q^{\text{con}} \right\}_{t+t_1} \right] \tag{3.13}$$

If we introduce the small parameter $\alpha = \sqrt{m_1/M}$, corresponding to the ratio of the thermal velocities of the Brownian particle and the molecules of the liquid, replace the momentum P by the reduced momentum $P^* = \alpha P$ (then $\partial/\partial P = \alpha \partial/\partial P^*$), and restrict the next analysis to the second order in α, we can make the following simplifications in equation (3.13):

a) represent the operator $U(t, t+t_1)$ in the zeroth approximation in α in the form $\exp(-iL_{fl})$, where L_{fl} is the Liouville operator of the fluid proper, since, in the zeroth approximation, the Brownian particle must be fixed;

b) analyze the term in square brackets in the same manner as was done in the integrand of equation (2.7.1) in the linear approximation in the velocities $\mathbf{v}^{\text{I}}$, $\mathbf{v}^{\text{S}}$, since they certainly do not exceed the velocity of the Brownian particle: $\mathbf{P}/M = \alpha \mathbf{P}^*$. This term will then give, on the right–hand side of equation (3.13), the sum $\Delta^1 + \Delta^2$, in terms of the type (2.7.14) and (2.7.15);

c) $f_B(t+t_1) = f_B(t)$;

d) assume on the right–hand side that $< \dot{\mathbf{P}} >_q^{\text{con}} \simeq < \dot{\mathbf{P}} >_0^{\text{con}} = 0$, where $< \ldots >_0^{\text{con}}$ denotes averaging with the conditional equilibrium distribution function of the liquid, with the Brownian particle at rest.

Then (3.13) takes the form

$$\frac{\partial f_B}{\partial t} + \frac{\mathbf{P}}{M}\frac{\partial f_b}{\partial \mathbf{R}} = \frac{\partial}{\partial \mathbf{P}} \left[-(< \dot{\mathbf{P}}(1+\Delta^1) >_q^{\text{con}} + < \dot{\mathbf{P}}\Delta^2 >_q^{\text{con}}) f_B + kT\zeta \frac{\partial f_B}{\partial \mathbf{P}} \right] \tag{3.14}$$

where

$$\zeta = \frac{1}{kT} \int_{-\infty}^{0} dt \, e^{\varepsilon t} < \dot{\mathbf{P}}\dot{\mathbf{P}}(t) >_q^{\text{con}} \tag{3.15}$$

is the drag coefficient of the Brownian particle.

To reduce the resulting kinetic equation to its final form, it is necessary to calculate the mean values on the right–hand side, including the coefficient ζ. Then the first problem we must consider is which forces determine the acceleration $\dot{\mathbf{P}}$ of the Brownian particle, and this depends on the behavior of the surface layer.

5.4 Boundary Conditions on the Surface of the Large Particle

In accordance with the investigation made in Chapter 2 of the boundary conditions on a rough solid surface, the velocity of the surface layer $\mathbf{v}^{\mathrm{S}}$ must be equal to the velocity of the particle surface $\mathbf{v}^{\mathrm{II}}$, since the coefficient of the friction ζ^{II} on the boundary Σ_{II} with the solid surface is much greater than the coefficient of friction ζ^{I} on the boundary Σ_{I} with the bulk liquid phase. In this case, slip, determined by equation (2.10.20), can occur on the boundary Σ_{I}. The correlation length l_0 in the liquid is very short and, therefore, the slip on Σ_{I} is also very small. Therefore, in this example we shall assume that there is no slip on the surface Σ_{I}. We must then have a no–slip stick boundary condition:

$$\mathbf{v}^{\mathrm{I}}|_{\Sigma_{\mathrm{I}}} = \mathbf{v}^{\mathrm{II}} \equiv \mathbf{P}/M \tag{4.1}$$

The last equation is valid for a nonrotating particle.

The stick boundary condition (4.1) provides a continuity of the velocity profile in the system. As a result, the motion equation for the momentum density takes the form of the conservation law (2.5.11). Integrating equation (2.5.11) over the volume of the Brownian particle we find that the acceleration of the Brownian particle is determined by the momentum flow through the surface Σ_{I}:

$$\dot{\mathbf{P}} = -\int d\mathbf{x}\, \Theta^{\mathrm{I}} \dot{\mathbf{p}}(\mathbf{x}) = \int_{\Sigma_{\mathrm{I}}} ds\, \mathbf{n}\mathbf{T}(\mathbf{x}) \tag{4.2}$$

and the term with Δ^2 on the right–hand side of equation (3.14) is proportional to the discontinuities of the velocity on the boundaries, and vanishes in their absence.

Note that in this section we have introduced a specific feature characteristic of the Brownian motion of a large particle in a liquid. For a large particle in a gas, it is necessary to modify the boundary condition (4.1) for $\mathbf{v}^{\mathrm{I}}$ to allow slip on this boundary, and correspondingly take into account terms with Δ^2 on the right–hand side of equation (3.14).

Another limiting case is a heavy impurity molecule. Since the linear dimension is of the order of the range of the intermolecular interaction, it can be regarded as a point particle that has no volume and surface Σ_{I}. Thus, it is unsuitable to be regarded as a bulk phase. The Brownian motion of such particles will be considered later.

5.5 Calculation of the Average Force and the Drag Coefficient

We now analyze the Brownian motion of a large particle in a liquid when the boundary condition (4.1) is satisfied and the force acting on the particle is determined by the expression (4.2). Then the kinetic equation (3.14) and the drag coefficient take the forms

$$\frac{\partial f_B}{\partial t} + \frac{\mathbf{P}}{M}\frac{\partial f_B}{\partial \mathbf{R}} = \frac{\partial}{\partial \mathbf{P}}\left[-\int ds\,\mathbf{n}\left\langle \mathbf{T}(\mathbf{x})(1+\Delta^1)\right\rangle_q' f_B + kT\zeta\frac{\partial f_B}{\partial \mathbf{P}}\right] \quad (5.1)$$

$$\zeta = \beta\int_\Sigma ds\int_\Sigma ds'\int_{-\infty}^0 dt\,\mathbf{nn} : \left\langle \mathbf{T}(\mathbf{x})(\mathbf{T}(\mathbf{x}',t) - \langle \mathbf{T}(\mathbf{x}',t)\rangle_0')\right\rangle_0' \quad (5.2)$$

To get a better insight into the further calculations of expressions (5.1) and (5.2), we shall first restrict our consideration to a simpler case of the Brownian motion in a homogeneous liquid. Then Δ^1 is only different from zero as a result of the liquid state perturbations by the moving particle.

The first term in the right–hand side of equation (5.1) has been discussed in detail in Section 2.8 of Chapter 2 for the general case of the surface layer. Here we deal with the simpler well–known situation for the bulk fluid phase, where this term may be represented as

$$\int_\Sigma ds\,\mathbf{n}\left\langle \mathbf{T}(\mathbf{x})(1+\Delta^1)\right\rangle_q^{\text{con}} = -\int_\Sigma ds\,\mathbf{n}\sigma(\mathbf{x}) \quad (5.3)$$

where $\sigma(\mathbf{x})$ is the macroscopic stress tensor of the fluid. For simplicity, we restrict ourselves to the approximation of a viscous incompressible liquid. Then $\sigma(\mathbf{x})$ has the form

$$\sigma_{\alpha\beta}(\mathbf{x}) = -p(\mathbf{x})\delta_{\alpha\beta} + \eta\left(\frac{\partial v_\alpha}{\partial x_\beta} + \frac{\partial v_\beta}{\partial x_\alpha}\right) \quad (5.4)$$

where, as before, $p(\mathbf{x}) = \langle \mathbf{T}(\mathbf{x})\rangle_q'$ is the pressure and η is the coefficient of viscosity of the bulk liquid, which can be represented in the form of the Kubo formula

$$\eta\left(\delta_{\alpha\mu}\delta_{\beta\nu} + \delta_{\alpha\nu}\delta_{\beta\mu} - \frac{2}{3}\delta_{\alpha\beta}\delta_{\mu\nu}\right) = \beta\int_{-\infty}^0 dt\,e^{\varepsilon t}\int d\mathbf{x}'\left\langle \mathring{T}_{\alpha\beta}(\mathbf{x})\mathring{T}_{\mu\nu}(\mathbf{x}',t)\right\rangle_q \quad (5.5)$$

Here the subscripts α, β, μ, and ν sweep over x, y, z.

To further calculate expression (5.3), it is necessary to find the velocity field $\mathbf{v}^{\mathrm{I}}(\mathbf{x},t)$ and substitute it into (5.4) to find the stress field. The velocity field is found by solving the Navier–Stokes equation obtained by substituting (5.4) in (2.5.28) with the boundary conditions (4.1) on the surface of the particle and at infinity, $\mathbf{v}^{\mathrm{I}}|_\infty = 0$. This problem has been studied in detail in hydrodynamics (see, for example, [56]). The solution leads to the expression

$$\int_\Sigma ds\, \mathbf{n}\sigma(\mathbf{x}) = \zeta_S \mathbf{P}/M \tag{5.6}$$

We now turn to the calculation of the drag coefficient (5.2). We use the device proposed by Fox and Uhlenbeck [47], for which we multiply equation (5.2) by the tensor $P_\alpha P_\beta/M^2$ and use Gauss's theorem in conjunction with condition (4.1):

$$\frac{P_\alpha P_\beta}{M^2}\zeta_{\alpha\beta} = \beta \int\limits_{-\infty}^{0} dt\, e^{\varepsilon t} \int d\mathbf{x} \int d\mathbf{x}' \left\langle \frac{\partial}{\partial x_\mu} v_\alpha(\mathbf{x}) T_{\alpha\mu}(\mathbf{x}) \frac{\partial}{\partial x'_\nu} v_\beta(\mathbf{x}') \right.$$
$$\left. \times \left(T_{\beta\nu}(\mathbf{x}', t) - \langle T_{\beta\nu}(\mathbf{x}')\rangle_0\right)\right\rangle_0$$

whence

$$\begin{aligned}\frac{P_\alpha P_\beta}{M^2}\zeta_{\alpha\beta} = \beta \int\limits_{-\infty}^{0} dt \int d\mathbf{x} \int d\mathbf{x}' \Bigg[& \frac{\partial v_\alpha(\mathbf{x})}{\partial x_\mu}\frac{\partial v_\beta(\mathbf{x}')}{\partial x'_\nu} \\ & + \frac{\partial v_\alpha(\mathbf{x})}{\partial x_\mu} v_\beta(\mathbf{x}) \frac{\partial}{\partial x'_\nu} + v_\alpha(\mathbf{x}) \frac{\partial v_\beta(\mathbf{x}')}{\partial x'_\nu}\frac{\partial}{\partial x_\mu} \\ & + v_\alpha(\mathbf{x}) v_\beta(\mathbf{x}') \frac{\partial^2}{\partial x_\mu \partial x'_\nu}\Bigg] \\ & \times \left\langle T_{\alpha\mu}(\mathbf{x}) \left(T_{\beta\nu}(\mathbf{x}', t) - \langle T_{\beta\nu}(\mathbf{x}', t)\rangle_q\right)\right\rangle_q \end{aligned} \tag{5.7}$$

In the local approximation, the correlation function $\langle \mathbf{T}(\mathbf{x})\mathbf{T}(\mathbf{x}', t)\rangle_q$ is supposed to decay with an increase of the difference $\mathbf{x}-\mathbf{x}'$ so quickly that the hydrodynamic velocity $\mathbf{v}(\mathbf{x}')$ can be set equal to $\mathbf{v}(\mathbf{x})$. Then in accordance with equation (5.5), the first term of the integrand in equation (5.7) is proportional to the viscosity coefficient η and the other terms – to its space derivatives, so they can be neglected.

We therefore get

$$\begin{aligned}\frac{P_\alpha P_\beta}{M^2}\zeta_{\alpha\beta} &= \eta \int d\mathbf{x}\, \frac{\partial v_\alpha(\mathbf{x})}{\partial x_\mu}\frac{\partial v_\beta(\mathbf{x})}{\partial x_\nu}(\delta_{\alpha\beta}\delta_{\mu\nu} + \delta_{\alpha\nu}\delta_{\mu\beta} - \frac{2}{3}\delta_{\alpha\mu}\delta_{\beta\nu}) \\ &= \eta \int d\mathbf{x} \left(\frac{\partial v_\alpha}{\partial x_\beta}\frac{\partial v_\beta}{\partial x_\alpha} + \frac{\partial v_\alpha}{\partial x_\beta}\frac{\partial v_\alpha}{\partial x_\beta}\right)\end{aligned} \tag{5.8}$$

where we have taken into account the condition of incompressibility $\partial v_\alpha / \partial x_\alpha = 0$. Using the equations

$$\frac{\partial^2}{\partial x_\alpha \partial x_\beta} v_\alpha v_\beta = \frac{\partial v_\beta}{\partial x_\alpha}\frac{\partial v_\alpha}{\partial x_\beta} \qquad \frac{\partial}{\partial x_\beta} v_\alpha \frac{\partial}{\partial x_\beta} = \frac{\partial v_\alpha}{\partial x_\beta}\frac{\partial v_\alpha}{\partial x_\beta} + v_\alpha \frac{\partial^2}{\partial x_\beta^2} v_\alpha \tag{5.9}$$

and Navier–Stokes equation in the Stokes approximation, that is, $\frac{\partial}{\partial x_\alpha}\sigma_{\alpha\beta} = 0$, we get equation (5.8), in the form

$$\begin{aligned}\frac{P_\alpha P_\beta}{M^2}\zeta_{\alpha\beta} &= \eta \int d\mathbf{x} \left(\frac{\partial}{\partial x_\beta} v_\alpha \frac{\partial v_\alpha}{\partial x_\beta} + \frac{\partial}{\partial x_\beta} v_\alpha \frac{\partial v_\beta}{\partial x_\alpha} - \frac{1}{\eta}\frac{\partial}{\partial x_\beta} v_\alpha p(\mathbf{x})\delta_{\alpha\beta} \right) \\ &= \frac{P_\alpha}{M} \int_\Sigma ds\, n_\beta \left[\left(\frac{\partial v_\alpha}{\partial x_\beta} + \frac{\partial v_\beta}{\partial x_\alpha} \right) - p\delta_{\alpha\beta} \right]\end{aligned} \tag{5.10}$$

Since P_α are arbitrary, it follows that

$$\frac{P_\beta}{M}\zeta_{\alpha\beta} = \int_\Sigma ds\, n_\beta \sigma_{\alpha\beta} \tag{5.11}$$

whence for an isotropic liquid in accordance with equation (5.6)

$$\zeta = \frac{1}{3}\,\mathrm{Sp}\,\zeta_{\alpha\beta} = \zeta_S$$

Thus, the final equation for a Brownian motion of a large particle in a homogeneous viscous incompressible liquid takes the standard form of the Fokker–Planck equation

$$\frac{\partial f_B}{\partial t} + \frac{\mathbf{P}}{M}\frac{\partial f_B}{\partial \mathbf{R}} = \zeta_S \frac{\partial}{\partial \mathbf{P}} \left(\frac{\mathbf{P}}{M} f_B + kT \frac{\partial f_B}{\partial \mathbf{P}} \right) \tag{5.12}$$

Now we discuss the more general case of Brownian motion in a hydrodynamically nonequilibrium liquid. The heat capacity of the particle is considered so small that disturbances of the temperature field introduced by the particle motion would be negligible.

We turn to the calculation of the first term on the right–hand side of equation (5.1). It is appropriate to recall that in accordance with the general approach developed in Chapter 2, the surface includes not only the solid particle but also the adjoining surface layer of fluid with a thickness of the order of the scale of the irregularities of the solid surface. Thus, the averaging of this term is done in the isotropic fluid, where the correlation functions of the tensor $\mathbf{T}(\mathbf{x})$ and the vectors $\mathbf{j}_Q(\mathbf{x})$ and $\mathbf{j}_\alpha(\mathbf{x})$ are zeros by virtue of Curie's principle. Therefore equations (5.3) through (5.5) still stand, but the pressure p and the velocity $\mathbf{v}$ in equation (5.4) are represented by

$$p(\mathbf{x},t) = p_1(\mathbf{x},t) + p_2(\mathbf{x},t)\,, \quad \mathbf{v}(\mathbf{x},t) = \mathbf{v}_1(\mathbf{x},t) + \mathbf{v}_2(\mathbf{x},t) \tag{5.13}$$

where p_1 and $\mathbf{v}_1$ characterize the liquid state undisturbed by the particle and p_2 and $\mathbf{v}_2$ are disturbances generated by the motion of the particle. Assuming that ∇p_1 remains constant over a distance of the order of the particle diameter, we obtain the Archimedean force

$$\int_\Sigma ds\, \mathbf{n} p_1(\mathbf{x},t) U = -\frac{4}{3}\pi r_0^3 \nabla p_1 \tag{5.14}$$

Note that the linear in $\nabla\mathbf{v}_1$ contribution to (5.3) is zero. Therefore, for the following calculation of the second term in (5.1) it is sufficient to find the velocity field $\mathbf{v}_2(\mathbf{x},t)$. It is found by solving the Navier–Stokes equation with the boundary condition (4.1) on the surface of the particle and at infinity: $\mathbf{v}_2|_\infty = 0$. The solution to this problem [56] in the Stokes approximation gives

$$\int ds\, \mathbf{n}\sigma(\mathbf{x},t) = \zeta_S(\mathbf{P}/M - \mathbf{v}_1(\mathbf{x},t)) \tag{5.15}$$

In calculating the drag coefficient, as before, we multiply (5.2) by the tensor $(P_\alpha - Mv_{1\alpha})(P_\beta - Mv_{1\beta})/M^2$. Then

$$\begin{aligned} \frac{(P_\alpha - Mv_{1\alpha})(P_\beta - Mv_{1\beta})}{M^2}\zeta_{\alpha\beta} &= \beta \int\limits_{-\infty}^{0} dt\, e^{\varepsilon t} \int d\mathbf{x} \int d\mathbf{x}' \\ &\times \left\langle \frac{\partial}{\partial x_\mu} v_{2\alpha}(\mathbf{x}) T_{\alpha\mu}(\mathbf{x}) \frac{\partial}{\partial x'_\nu} v_{2\beta}(\mathbf{x}') \left(T_{\beta\nu}(\mathbf{x}',t) - \langle T_{\beta\nu}\rangle_0^{\rm con}\right) \right\rangle^{\rm con} \end{aligned} \tag{5.16}$$

In the local approximation we then find

$$\frac{P_\beta - Mv_{1\beta}}{M}\zeta_{\alpha\beta} = \int_\Sigma ds\, n_\beta \sigma_{\alpha\beta} \tag{5.17}$$

from which, as before, it follows that $\zeta \equiv \mathrm{Sp}\,\zeta_{\alpha\beta}/3 = \zeta_S \equiv 6\pi\eta r_0$.

Thus, the kinetic equation for a large spherical Brownian particle in an inhomogeneous fluid takes the form

$$\frac{\partial f}{\partial t} + \frac{\mathbf{P}}{M}\frac{\partial f}{\partial \mathbf{R}} - \frac{4}{3}\pi r_0^3 \nabla p_1 \frac{\partial f}{\partial \mathbf{P}} = \zeta_S \frac{\partial}{\partial \mathbf{P}}\left[\left(\frac{\mathbf{P}}{M} - \mathbf{v}_1(\mathbf{R},t)\right) + \frac{1}{\beta}\frac{\partial f}{\partial \mathbf{P}}\right] \tag{5.18}$$

It differs from an ordinary Fokker–Planck equation only in the addition of the term representing the Archimedean force on the left–hand side and the appearance of the velocity of the fluid on the right.

5.6 Brownian Diffusion of Large Particle in Inhomogeneous Fluid

On the basis of the obtained Fokker–Planck equation (5.18), it is now easy to derive a diffusion equation for Brownian particles.

We consider a system of noninteracting N Brownian particles such that their individual distribution functions satisfy the Fokker–Planck equation (5.18). Multiplying this equation by N and integrating with respect to the momentum $\mathbf{P}$, we get the first moment equation

$$\frac{\partial u(R,t)}{\partial t} + \frac{\partial}{\partial \mathbf{R}} \mathbf{n}(\mathbf{R},t)\mathbf{V}(\mathbf{R},t) = 0 \tag{6.1}$$

where $n(\mathbf{R},t) = N \int d\mathbf{P}\, f_B(\mathbf{P},\mathbf{R},t)$ is the mean density of the particles and

$$V(R,t) = \frac{N}{n(\mathbf{R},t)} \int dP \frac{\mathbf{P}}{M} f_B(\mathbf{P},\mathbf{R},t) \tag{6.2}$$

is the mean velocity of the particles.

Equation (6.1) is not closed, since it contains the velocity V. To close it we use the next moment equation that is the result of multiplication of equation (5.8) by $N\mathbf{P}$ and subsequent integration with respect to $\mathbf{P}$:

$$\begin{aligned} M\frac{\partial}{\partial t}\left(n(\mathbf{R},t)\mathbf{V}(\mathbf{R},t)\right) + \frac{\partial}{\partial \mathbf{R}}\left(n(\mathbf{R},t)kT(\mathbf{R},t)\right) + \frac{4}{3}\pi r_0^3 n(\mathbf{R},t)\nabla p_1 \\ = -\zeta_S n(\mathbf{R},t)\left(\mathbf{V}(\mathbf{R},t) - \mathbf{v}_1(\mathbf{R},t)\right) \end{aligned} \tag{6.3}$$

where

$$n(\mathbf{R},t)kT(\mathbf{R},t)U = \int d\mathbf{P}\, \frac{\mathbf{PP}}{M} f_B(\mathbf{P},\mathbf{R},t) \tag{6.4}$$

is the particular pressure tensor of the Brownian particles. Here the temperature $T(\mathbf{R},t)$ corresponds to the mean kinetic energy of translational motion of the particles. If we restrict the discussion to the case of relatively slow processes, the temperature $T(\mathbf{R},t)$ can be seen as equal to the thermodynamic temperature of the fluid at the point $\mathbf{R}$. As the letter is known this assumption terminates the hierarchy of moment equations: now equations (6.1) and (6.3) are a closed set of two equations in two unknown functions $n(\mathbf{R},t)$ and $\mathbf{V}(\mathbf{R},t)$. Hence in the steady approximation of equation (6.3) we obtain the diffusion flow in the form

$$\mathbf{I}_{diff} \equiv n(\mathbf{R},t)\left(\mathbf{V}(\mathbf{R},t) - \mathbf{v}_1(\mathbf{R},t)\right) = -D\nabla n - D_T\frac{\nabla n}{T} - D_p\nabla p_1 \tag{6.5}$$

For the coefficients of diffusion D, thermodiffusion, D_T, and barodiffusion, D_p, the following expressions are found:

$$D = kT/\zeta_S \tag{6.6}$$

$$D_T = nkT/\zeta_S = nD \tag{6.7}$$

$$D_p = \varphi/\zeta_S = \frac{\varphi}{kT}D, \quad \varphi = \frac{4}{3}\pi a^3 n \tag{6.8}$$

The first of them is the well-known Einstein relation. We postpone our discussion of these expressions until Section 5.10.

5.7 On a Limitation of the Theory

Note that in deriving equation (5.12) we have used the assumption that the flow of the liquid near the particle is stationary. In hydrodynamics, there is [56] the stationarity criterion $r_0^2 \ll \eta\tau/\rho$, where r_0 is the characteristic dimension (here, the radius of the particle), τ is the characteristic time, and ρ is the density of the liquid. For Brownian motion of a large particle $\tau \sim M/\zeta_S = \frac{2}{9}r_0^2\rho_B/\eta$, where ρ_B is the density of the matter of which the Brownian particle is composed. The final stationarity condition takes the form

$$\rho_B/\rho \gg 9/2 \tag{7.1}$$

i.e., the density ρ_B must exceed the density of the liquid by at least an order of magnitude.

If this is not the case, one cannot calculate the drag of the moving particle in the stationary approximation. For nonstationary motion, we obtain, instead of the Stokes formula, the Boussinesq formula for the drag of an arbitrarily moving sphere:

$$\int_\Sigma ds\, \mathbf{n}\sigma(\mathbf{x},t) = \frac{3}{2}\frac{\rho}{\rho_B}\left\{\frac{1}{3}\dot{\mathbf{P}} + \frac{3\eta}{\rho r_0^2}\mathbf{P} + \frac{3}{r_0}\sqrt{\frac{\eta}{\pi\rho}}\int_{-\infty}^{t}\frac{d\tau}{\sqrt{t-\tau}}\dot{\mathbf{P}}(\tau)\right\} \tag{7.2}$$

A similar expression will be obtained in the nonstationary case for the calculation of (5.2). Since the right-hand side of equation (5.1) will now depend on the acceleration $\dot{\mathbf{P}}$ and contain the non-Markov term that takes into account the retardation in the third term in (7.2), equation (5.1) cannot be reduced to the Fokker-Planck form (5.12).

Therefore, the inequality (7.1) is not merely the criterion of admissibility of the stationary Stokes approximation for the drag, but also the condition for Markov Brownian motion of a large spherical particle, as it permits time-scale separation between fluid and particle velocities underlying the Fokker-Planck equation.

5.8 Derivation of the Fokker–Planck Equation for the Point Brownian Particle

Now we return to another limiting case of a heavy impurity molecule. Recall that generality of our analysis was lost in Section 5.4 when formulating the boundary conditions at the surface of the Brownian particle. As an impurity molecule can be regarded as a point particle, it has no surface and its motion does not give rise to hydrodynamic perturbations in the fluid. So the general structure of equation (2.7.14) for Δ^1 is not changed when introducing the heavy point particle into the hydrodynamically nonequilibrium fluid. It is necessary here to note that the equation (2.7.14) has been derived from equation (2.7.2) using the conservation laws (2.5.11) and (2.5.13). In the case of a point impurity we are dealing with an extremely degenerate example of the two–phase system (namely, one of two phases becomes the point), so we cannot use the procedure accepted to transform equation (2.5.8) into the conservation law (2.5.11). Thus, the momentum conservation law is to be used in the form (2.5.8):

$$\dot{p}(\mathbf{x}) = -\operatorname{Div}\mathbf{T}'(\mathbf{x}) + \mathbf{F}(\mathbf{x}) \tag{8.1}$$

The energy conservation law has a source term in its right–hand side that may not be represented as divergence of a vector:

$$\dot{H}(\mathbf{x}) = -\operatorname{div}\mathbf{j}'_H(\mathbf{x}) + J(\mathbf{x}) \tag{8.2}$$

An explicit expression for the source term $J(\mathbf{x})$ is irrelevant for the subsequent discussion. Suffice it to note that it is a scalar.

The presence of the source terms in the conservation laws for momentum and energy leads to additional terms in equation (2.7.14); they are

$$\begin{aligned}\Delta_F = &-\int d\mathbf{x}\int_{-\infty}^{0} dt_1 \sum_l \Theta^l(\mathbf{x},t+t_1)\beta^l(\mathbf{x},t+t_1)\mathbf{v}^l(\mathbf{x},t+t_1)\\ &\times\sum_{\alpha i}\frac{\partial u^{0\alpha}(\mathbf{R}(t_1)-\mathbf{r}_{\alpha i}(t_1))}{\partial\mathbf{R}(t_1)}\left[\delta(\mathbf{x}-\mathbf{R}(t_1))-\delta(\mathbf{x}-\mathbf{r}_{\alpha i}(t_1))\right]\end{aligned} \tag{8.3}$$

$$\Delta_J = \int d\mathbf{x}\int_{-\infty}^{0} dt_1 \sum_l \Theta^l(\mathbf{x},t+t_1)\beta^l(\mathbf{x},t+t_1)J(\mathbf{x},t_1) \tag{8.4}$$

Note that equation (2.3) for the center of mass dynamic density transforms into the one–particle dynamic density; the projection operator (2.5) specified by this density will therefore determine the one–particle distribution function when acting as in equation (2.4) on the full distribution function.

As the point Brownian particle does not carry away the fluid, and its velocity is P/M, equation (8.3) can be integrated in $\mathbf{x}$ and rewritten in the form

$$\Delta_F = \beta(\mathbf{R},t)\left(\frac{\mathbf{P}}{M} - \mathbf{v}(\mathbf{R},t)\right)\int_{-\infty}^{0} dt_1\, \mathbf{F}(\mathbf{R},t) \tag{8.5}$$

where

$$\mathbf{F}(\mathbf{R}) = -\sum_{\alpha i}\frac{\partial u^{0\alpha}(\mathbf{R}-\mathbf{r}_{\alpha i})}{\partial \mathbf{R}} \tag{8.6}$$

It is necessary to note here that the acceleration of the point particle, $\dot{\mathbf{P}} = \mathbf{F}(\mathbf{R})$, is determined by the vector dynamic value (but not the divergence of the tensor value, as in the case of a large particle). So in accordance with the Curie principle only vector terms in the sum $(\Delta^1+\Delta_F+\Delta_J)$ are taken into account when calculating the mean $\left\langle \dot{\mathbf{P}}(\Delta^1+\Delta_F+\Delta_J)\right\rangle_0$. They are the term Δ_F and the flows of energy and particles numbers in the term Δ^1. Thus equations (2.14) and (3.15) take the forms

$$\begin{aligned}\frac{\partial f_B}{\partial t} + \frac{\mathbf{P}}{M}\frac{\partial f_B}{\partial \mathbf{R}} + \langle \mathbf{F}\rangle_q'\frac{\partial f_B}{\partial \mathbf{P}} = \frac{\partial}{\partial \mathbf{P}}\Bigg\{\zeta_k\left[\left(\frac{\mathbf{P}}{M} - \mathbf{v}(\mathbf{R},t)\right) f_B + kT\frac{\partial f_B}{\partial \mathbf{P}}\right] \\ -\eta\frac{\nabla\beta}{\beta} f_B - \sum_\alpha \chi_\alpha \frac{\nabla \nu_\alpha}{\beta} f_B\Bigg\}\end{aligned} \tag{8.7}$$

$$\zeta_k = \beta(\mathbf{R},t)\int_{-\infty}^{0} dt\, \langle \mathbf{F}(\mathbf{R})\mathbf{F}(\mathbf{R},t)\rangle_0' \tag{8.8}$$

There are two additional kinetic coefficients in equation (8.7):

$$\eta = \beta(\mathbf{R},t)\int_{-\infty}^{0} dt\, \langle \mathbf{F}(\mathbf{R})\mathbf{j}_H(\mathbf{R},t)\rangle_0' \tag{8.9}$$

is the coefficient of the thermal drag of the Brownian particle by the fluid energy flow (thermophoresis coefficient), and

$$\chi_\alpha = \beta(\mathbf{R},t)\int_{-\infty}^{0} dt\, \langle \mathbf{F}(\mathbf{R})\mathbf{j}_\alpha(\mathbf{R},t)\rangle_0' \tag{8.10}$$

is the coefficient of the diffusion drag by the diffusion flow of the α component of the fluid (diffusiophoresis coefficient).

The mean force $\langle \mathbf{F} \rangle_q'$ on the left–hand side of equation (8.7) can be expressed in terms of the pair correlations function $g_\alpha(\mathbf{R}, \mathbf{r}_\alpha, t)$ that determines the probability of finding a molecule of the α–component of the fluid at a point $\mathbf{r}_\alpha$ under the condition of the Brownian particle presence at the point $\mathbf{R}$:

$$\langle \mathbf{F}(\mathbf{R}) \rangle_q' = \sum_\alpha \int d\mathbf{r}_\alpha \frac{\partial u^{0\alpha}(\mathbf{R} - \mathbf{r}_\alpha)}{\partial \mathbf{r}_\alpha} g_\alpha(\mathbf{R}, \mathbf{r}_\alpha, t) \tag{8.11}$$

It can be considered in a gas that

$$g_\alpha(\mathbf{R}, \mathbf{r}_\alpha, t) = n_\alpha(\mathbf{R}, t) \exp\{-\beta(\mathbf{r}_\alpha, t) u^{0\alpha}\} \tag{8.12}$$

where $n_\alpha(\mathbf{R}, t)$ is the mean density of the α–component. Then, in the linear approximation in gradients equation (8.11) takes the form

$$\langle \mathbf{F} \rangle_q' = -\sum_\alpha \frac{4}{3}\pi a_{0\alpha}^3 \nabla p_\alpha + b_0 \frac{\nabla T}{T} \tag{8.13}$$

where $a_{0\alpha}$ is an effective radius of the particle determined via its volume as

$$\frac{4}{3}\pi a_{0\alpha}^3 = \int d\mathbf{r}_\alpha \left(1 - e^{-\beta u^{0\alpha}}\right)$$

and $p_\alpha(\mathbf{R}, t) = n_\alpha(\mathbf{R}, t)/\beta(\mathbf{R}, t)$ is the partial pressure, and

$$b_0 = \sum_\alpha \int d\mathbf{r}_\alpha g_\alpha(\mathbf{R}, \mathbf{r}_\alpha, t) u^{0\alpha}(\mathbf{R} - \mathbf{r}_\alpha)$$

If the potential $u^{0\alpha}$ is the same hard sphere potential for all α then $a_{0\alpha} = a_0$ and $b_0 = 0$ for all α. In this case the mean force (8.13) becomes the Archimedean force:

$$\langle \mathbf{F} \rangle_q' = -\frac{4}{3}\pi a_0^3 \nabla p \tag{8.14}$$

where $p = \sum_\alpha p_\alpha$ is the full gas pressure.

The friction coefficient was first derived in such a form, (8.8), by Kirkwood [49]. The thermal drag coefficient (8.9) was first obtained by Zubarev and the author [57, 58] and the diffusion drag coefficient was first obtained by Mazo [59].

It is interesting to note that the nonequilibrium fluid state, on the one hand, gives rise to a sort of an external force field $\langle \mathbf{F} \rangle_q'$ and, on the other hand, results in the appearance of additional dissipative effects due to the drag of the Brownian particle by the heat and diffusion flows in the system, in contrast to the previous case for a large Brownian particle. There the drag coefficient η and χ_α were equal to zeros as the force of the large particle interaction with the fluid is of a tensor nature, and in accordance to the Curie principle, its correlation functions with the vectors of the heat and diffusion flows vanish.

5.9 Kirkwood Friction Coefficient

It was shown in Section 5.6 that in the case of a large Brownian particle its friction coefficient is described by the well–known Stokes formula. It is interesting to evaluate expression (8.8) and compare it with the previous result.

For brevity we confine the analysis to the first approximation in the fluid density n. Then equation (8.8) can be rewritten as

$$\zeta_k = \beta(\mathbf{R}, t) n \int_{-\infty}^{0} dt \int d\mathbf{x}_1 \, \frac{\partial u}{\partial \mathbf{r}_1} e^{i\mathcal{L}_1 t} \frac{\partial u}{\partial \mathbf{r}_1} f_1^{(0)}(\mathbf{x}) \tag{9.1}$$

where $\mathbf{x}_1 = (\mathbf{p}_1, \mathbf{r}_1)$ is the phase coordinate of a gas molecule, $f_1^{(0)}(\mathbf{x})$ is the one–particle equilibrium gas distribution, and $\mathcal{L}_1$ is the Liouville operator of a gas molecule in the force field of the Brownian particle described by the one–particle Hamiltonian

$$H_1 = \frac{p^2}{2m} + u(\mathbf{r} - \mathbf{R}) \tag{9.2}$$

Since, in such a system,

$$\dot{\mathbf{p}} = i\mathcal{L}_1 \mathbf{p} = -\frac{\partial u}{\partial \mathbf{r}} \tag{9.3}$$

the integration with respect to time, t, in equation (9.1) can be easyly performed, and we get

$$\int_{-\infty}^{0} dt \, e^{i\mathcal{L}_1 t} \frac{\partial u}{\partial \mathbf{r}_1} = \mathbf{p}(t = 0) - \mathbf{p}(t = -\infty) \tag{9.4}$$

As in the lowest approximation in the gas density the intermolecular interaction in the gas can be neglected. The molecule motion in the central force field, $u(\mathbf{r} - \mathbf{R})$, is symmetrical with respect to time, so

$$\mathbf{p}(t = 0) - \mathbf{p}(t = -\infty) = \frac{1}{2} \left[\mathbf{p}(t = +\infty) - \mathbf{p}(t = -\infty)\right] \tag{9.5}$$

We present the right–hand side of this equation in the form

$$\mathbf{p}(t = +\infty) - \mathbf{p}(t = -\infty) = 2(\mathbf{p}_\infty \cdot \mathbf{k})\mathbf{k} \tag{9.6}$$

where $\mathbf{k}$ is the unit vector directed along the normal to the Brownian particle surface at the point of collision so that

$$\frac{\partial u}{\partial \mathbf{r}} = \mathbf{k}\frac{\partial u}{\partial r} \tag{9.7}$$

To determine the force $\partial u/\partial r$ for the case of the Brownian particle being the hard sphere with the specular reflections of colliding gas molecules, we consider the auxiliary problem of the one–dimensional motion of the molecule with reflection from an elastic wall. The solution of this problem is evident: at an initial time $t = 0$ the molecule moves freely with the velocity v from the point r_1 to the wall at $r = r_0$ ($r_0 < r_1$). Then, at the time $t = (r_1 - r_0)/v$, it reflects from the wall and moves in the reverse direction. The velocity of the molecule can be written in the form

$$\dot{r}(t) = v\left[2\Theta\left(vt - (r_1 - r_0)\right) - 1\right] \tag{9.8}$$

where Θ is the Heaviside function of unit step. Thus, we get

$$-\frac{\partial u}{\partial r} = m\ddot{r} = 2\frac{p^2}{m}\delta\left(vt - (r_1 - r_0)\right) \tag{9.9}$$

or

$$\frac{\partial u}{\partial r} = -2\frac{p^2}{m}\delta(r - r_0) \tag{9.10}$$

since the interaction of the molecule with the Brownian particle occurs at $t = 0$. True, such an interpretation of the interaction force $\partial u/\partial r$ replace the problem of a point heavy impurity in a liquid by the problem of a finite sphere in a rarefied gas. This substitution seems to be well suited as it conserves the main distinctive properties of the starting problem: 1) the interaction between the particle and the fluid is described by the spherically symmetric potential, and 2) the particle size is far less than the correlation length in the fluid.

Thus, introducing the scalar product (as it was done, for example, in equation (2.8.5)) with the use of equations (9.4) through (9.7) and (9.10), we get equation (9.1) in the form of the integral with respect to $d\mathbf{x} = d\mathbf{p}d\mathbf{r}$:

$$\begin{aligned}\zeta^{(0)} &= \frac{n}{3kT}\frac{1}{(2\pi mkT)^{3/2}}\int_0^{2\pi} d\Phi \int_0^{\pi} d\Theta \sin\Theta \int_{r_0}^{\infty} dr\, r^2 \int_0^{2\pi} d\varphi \int_0^{\pi} d\vartheta \sin\vartheta \\ &\times \int_0^{\infty} dp\, p^2 2\frac{p^2}{m}\delta(r - r_0)e^{-p^2/2mkT} p\cos\vartheta = \frac{16}{3}nkT\pi r_0^2\sqrt{\frac{m}{2\pi kT}}\end{aligned} \tag{9.11}$$

Thus, we have derived the well-known Epstein formula for the drag of the sphere moving in a rarefied gas.

Notice that equation (9.11) differs significantly from the Stokes formula by the quadratic dependence on r_0. Taking into account the next terms in the density expansion do not remove this contradiction [60]. To obtain the linear dependence of ζ on r_0, it is necessary to take into account perturbation of the fluid by the moving particle. There were [54, 61] semiphenomenological calculations of the Kirkwood formula taking account of such a perturbation by the particle whose interaction with the fluid molecules was described by a spherical potential. Since the type of interaction corresponds to a slip boundary condition (as on the bubble surface), this approach led to the relevant formula $\zeta = 4\pi\eta a_0$.

5.10 Diffusion of the Point Particle

Here we shall derive the diffusion equations for the point Brownian particles on the basis of the Fokker–Planck equation (8.7).

Following the general schedule used in Section 5.6 we obtain here the same first moment equation for the mean number density of the Brownian particles $n(\mathbf{R}, t)$. As for the second moment equation, there are additional terms with the coefficients η and χ_α. As a result the diffusion flow equation takes the form

$$\begin{aligned}\mathbf{I}_{diff} &\equiv n(\mathbf{x}, t)\,(\mathbf{V}\mathbf{x}, t - \mathbf{v}(\mathbf{x}, t)) \\ &= -D\nabla n - D_T\frac{\nabla T}{T} + \sum_\alpha D_\alpha \nabla \nu_\alpha + \frac{n}{\zeta_k}\left\langle \mathbf{F}(\mathbf{x})\right\rangle_q'\end{aligned} \tag{10.1}$$

where

$$D = kT/\zeta_k \tag{10.2}$$

$$D_T = nkT(1 + \eta/kT)/\zeta_k \tag{10.3}$$

$$D_\alpha = nkT\chi_\alpha/\zeta_k \tag{10.4}$$

In general the mean force term $n\left\langle \mathbf{F}\right\rangle_q' /\zeta_k$ is a linear form relative to the gradients ∇T and $\nabla\nu_\alpha$ and therefore it gives additional contributions to the coefficients D_T and D_α.

In the particular case of the hard spherical Brownian particle, the mean force term does not contribute to D_T and D_α. From equation (8.13), its influence reduces completely to one more parameter, namely, the pressure gradient responsible for the complementary effect of barodiffusion. Then equation (10.1) becomes

$$\mathbf{I}_{diff} = -D\nabla n - D_T \frac{\nabla T}{T} + \sum_{\alpha} D_{\alpha} \nabla \nu_{\alpha} - D_p \nabla p \qquad (10.5)$$

where

$$D_p = \varphi/\zeta_k = D\varphi/kT \qquad (10.6)$$

is the barodiffusion coefficient, and

$$\varphi = \frac{4}{3}\pi a^3 n \qquad (10.7)$$

is is the volume concentration of the Brownian particles.

Equation (10.2) is the well–known Einstein relation for the diffusion coefficient of Brownian particles, and equations (10.3) and (10.4) are the similar relations between the thermodiffusion and cross diffusion coefficients on one hand and the coefficients of thermophoresis and diffusiophoresis on the other.

Equation (10.3) was first obtained by Nicolis [62] and Zubarev and Bashkirov [57, 63]. Later it was shown [58] that the correct use of the Luttinger method [64] led to the same relation (10.3).

The barodiffusion coefficient (10.6) coincides with the Ramshow result [65] for this case.

Notice the distinctions between these equations and the related equations for large Brownian particles in Section 5.6. The main difference is a consequence of the additional terms in equation (8.7) in comparison with the equation (5.18).

Chapter 6

Brownian Motion of N Spheres in Inhomogeneous Fluids

6.1 Introduction

Until recently systems of large particles suspended in a liquid and interacting with each other via hydrodynamic viscous stresses in the liquid were studied mainly with using phenomenological approaches. They are based on approximate solutions to hydrodynamic equations with relevant boundary conditions on all particles and subsequent averaging over the particles' positions. It is impossible to apply directly the statistical mechanics methods to average the motion equations of particles with dissipative viscous interparticle interaction. We must accept some assumptions, therefore, that are ocasionally contradictory, which was demonstrated, for example, by a controversial discussion in [66, 67] arisen some years ago about a determination of the thermodynamic driving force of the diffusion of Brownian spheres in a viscous liquid.

In Chapter 5, a new statistical mechanical approach to the theory of Brownian motion of a single large particle in a fluid was developed. The explicit account of viscous stresses in the fluid produced by a moving particle is a principal advantage of this approach. Thus, an innovative approach is developed for a statistical theory of many–particle Brownian motion, taking into account hydrodynamic interactions, every particle in motion in the liquid being disturbed by the motion of other particles.

In Section 6.2 of this chapter the Fokker–Planck equation will be derived for a distribution function of centers of masses of N large Brownian particles, suspended in a hydrodynamically nonequilibrium viscous liquid, starting from the Liouville equation for the assembly of all molecules of the system (the liquid and the particles). The resultant Fokker–Planck equation takes into account for both potential and hydrodynamic nterparticle interactions.

In Section 6.3 we will derive the Smoluchowski equation for an N–particle coordinate distribution function. On this basis in Section 6.4, the diffusion equation will be found and relevant diffusion, thermodiffusion, and barodiffusion coefficients will be calculated.

In Section 6.5 the convection diffusion will be derived for the particular case of one Brownian particle in a liquid flowing through a grid of fixed spherical particles randomly distributed in space.

In Section 6.6 this equation is analyzed as a stochastic equation with a fluctuating convective velocity field. The diffusion coefficient of the suspended particle in such fluctuating media is calculated in Section 6.6, taking into account the influence of fixed particles.

6.2 The N-particle Fokker–Planck Equation

As in the case of one Brownian particle, here we will consider each particle to be an assembly of molecules of one type and the liquid to be an assembly of molecules of another type. Since we are not interested in intrinsic processes occuring within particles, the use of the following molecular description appears unnecessary complicated. The main point is that only in terms of this microscopic description all interactions in the system can be described by intermolecular potentials. Thus the Liouville theorem applies, and the Liouville equation can be written for the entire assembly of molecules.

This treatment does not neglect hydrodynamic interparticle interactions. Such an interaction appears as a result of transition to an averaged hydrodynamic description of a liquid — much as in kinetic gas theory — the viscous stresses, the heat and diffusion flows appear when passing from the kinetic description, using the Boltzmann equation, to the hydrodynamic one.

The Hamiltonian of the system consisting of the liquid and the suspended large particles can be written in the form

$$\mathcal{H} = \sum_i \left(\frac{p_i^2}{2m_1} + \frac{1}{2} \sum_j u_{ij} \right) + \sum_{K,k} \left(\frac{p_{Kk}^2}{2m_2} + \frac{1}{2} \sum_l \varphi_{kl}^K \right) + \sum_{Kil} \mathbf{V}_{il}^K + \frac{1}{2} \sum_{KLkl} \Phi_{kl}^{KL} \tag{2.1}$$

where indices i and j are used for the coordinates and momenta of the liquid molecules; k and l for the particle molecules; and K and L for the particles themselves. Next are the interaction potentials in equation (2.1): u_{ij} — between the ith and jth molecules of the liquid, φ_{kl}^K — between the kth and lth molecules of the Kth particle, Φ_{kl}^{KL} — between the kth molecule of the Kth particle and the lth molecule of the Lth particle. Notice that the potential φ_{kl}^K should have a sufficiently deep negative (attractive) component to provide the bound state of molecules forming the Kth particle within the considered temperature range.

The use of the molecular description of the system, with only potential interactions as a starting point makes it possible to use the Liouville equation for a nonequilibrium distribution function of all molecules of the system $\rho(\{\mathbf{p}_i\mathbf{r}_i\}, \{\mathbf{p}_k\mathbf{r}_k\}, t)$ as the basis to develop new theory.

Since any processes taking place within particles are beyond the scope of the present theory, we shall describe the particles using their center of mass coordinates, $\mathbf{R}_k$, and momenta, $\mathbf{P}_k$. With this in mind, the N-particle dynamical density is determined as the product of one–particle dynamical densities, equation (5.2.3):

$$n(\{\mathbf{P}_k, \mathbf{R}_k\}) = \prod_k \delta(\mathbf{P}_k - \sum_k \mathbf{p}_{kk})\, \delta(N_k\mathbf{R}_k - \sum_k \mathbf{r}_{kk}) \,. \tag{2.2}$$

The average of this density over the whole phase space with the total distribution function $\rho(\{\mathbf{p}_i\mathbf{r}_i\}, \{\mathbf{p}_k\mathbf{r}_k\}, t)$ is the N-particle distribution function of centers of mass coordinates and momenta of all large particles in the form

$$f^N(\{\mathbf{P}_k, \mathbf{R}_k\}, t) = \langle n(\{\mathbf{P}_k, \mathbf{R}_k\}) \rangle \ . \tag{2.3}$$

With the use of the projection operator

$$\mathcal{P}^N A = \rho_q'(t)\, \mathrm{Sp}[n(\{P_k, R_k\})A] \tag{2.4}$$

similar to one used in Section 5.2, the kinetic equation for f^N is derived from the relevant Liouville equation in the lowest approximation in the small parameter $\alpha = \sqrt{m_1/M}$. As the derivation is absolutely analogous to the derivation of equation (5.5.1), it is written as

$$\frac{\partial f^N}{\partial t} + \sum_K \left(\frac{\mathbf{P}_K}{M} \frac{\partial f^N}{\partial \mathbf{R}_K} + \frac{\partial}{\partial \mathbf{P}_K} \left\langle \dot{\mathbf{P}}_K \right\rangle_q' f^N \right)$$
$$= \sum_K \frac{\partial}{\partial \mathbf{P}_K} \left[\left\langle \dot{\mathbf{P}}_K \Delta \right\rangle_q' f^N + \sum_L \zeta_{KL} \frac{\partial f^N}{\partial \mathbf{P}_L} \right] \tag{2.5}$$

where

$$\zeta_{KL} = \beta \int_{-\infty}^{0} dt \left\langle \dot{\mathbf{P}}_K \left(\dot{\mathbf{P}}_L(t) - \left\langle \dot{\mathbf{P}}_L(t) \right\rangle_0 \right) \right\rangle_0 \tag{2.6}$$

is the tensor, its diagonal elements ζ_{KK} are the friction coefficients of the Kth particle, and nondiagonal elements describe interactions between different particles.

Notice that as in deriving equation (5.5.1), we have $\left\langle \dot{\mathbf{P}}_K \right\rangle_q' \simeq \left\langle \dot{\mathbf{P}}_K \right\rangle_0'$ in the zeroth approximation in α, but now this average is nonzero. Let us consider forces that give rise to the acceleration. In the former case of a single large particle, as shown in Section 5.4, the acceleration is determined entirely by the momentum flow across the particle surface.

In the N-particle case it is necessary to account for the forces $\mathbf{F}_{KL}$ of interaction between large particles due to the potential Φ^{KL} in the Hamiltonian (2.1).

Using the relative coordinates $\mathbf{r}'_{Kk} = \mathbf{r}_{Kk} - \mathbf{R}_K$, the forces of interaction between molecules possessed by different particles can be written as

$$\sum_{kl} \frac{\partial}{\partial \mathbf{r}_{Kk}} \Phi^{KL}(\mathbf{r}_{Kk} - \mathbf{r}_{Ll}) = \frac{\partial}{\partial \mathbf{R}_K} \sum_{kl} \Phi^{KL}(\mathbf{R}_K + \mathbf{r}'_{Kk} - \mathbf{R}_L - \mathbf{r}'_{Ll}) \tag{2.7}$$

If particles are considered to be hard spheres, the effective potential of the pair interparticle interaction can be introduced in the form

$$\tilde{\Phi}^{KL}(\mathbf{R}_K - \mathbf{R}_L) = \sum_{kl} \Phi^{KL}(\mathbf{R}_K + \mathbf{r}'_{Kk} - \mathbf{R}_L - \mathbf{r}'_{Ll})$$
$$\simeq \frac{N_K N_L}{\omega_K \omega_L} \int_{\omega_K} d\mathbf{r}'_K \int_{\omega_L} d\mathbf{r}'_L \, \Phi^{KL}(\mathbf{R}_K + \mathbf{r}'_K - \mathbf{R}_L - \mathbf{r}'_L) \tag{2.8}$$

where ω_K and ω_L are the volumes of the Kth and Lth particles.

Thus, the acceleration $\dot{\mathbf{P}}_K$ is determined to be

$$\dot{\mathbf{P}}_K = \mathcal{F}_K + \mathbf{F}_K \tag{2.9}$$

where

$$\mathcal{F}_K = \int_\Sigma ds\, \mathbf{n} T(\mathbf{x}) \tag{2.10}$$

is the momentum flow across the particle surface and

$$\mathbf{F}_K = \sum_L \mathbf{F}_{KL} = -\frac{\partial}{\partial \mathbf{R}_K} \sum_L \tilde{\Phi}^{KL}(\mathbf{R}_{KL}) \tag{2.11}$$

is the force of the elastic (potential) interaction of the Kth particle with other suspended particles.

As the potential $\tilde{\Phi}^{KL}$ depends only on the center of mass coordinates, the average acceleration (2.9) takes the form

$$\left\langle \dot{\mathbf{P}}_K \right\rangle' = \langle \mathcal{F}_K \rangle'_q + \mathbf{F}_K \tag{2.12}$$

where

$$\langle \mathcal{F}_K \rangle'_q = \int_{\Sigma_K} ds\, \mathbf{n}\, (p_1(\mathbf{x}, t) + p_2(\mathbf{x}, t)) = \mathbf{F}_K^A + \int_{\Sigma_K} ds\, \mathbf{n} p_2(\mathbf{x}, t) \tag{2.13}$$

$p_1(\mathbf{x}, t)$ is the pressure of the liquid nondisturbed by the particles, $p_2(\mathbf{x}, t)$ is the pressure due to the motion of the particles in the liquid, and $F_K^A = -\omega \nabla p_1$ is the Archimedean force acting on the particle with the volume $\omega = \frac{4}{3}\pi a^3$ immersed in the liquid with the proper pressure gradient ∇p_1.

Thus, equation (2.5) can be rewritten in the form

$$\frac{\partial f^N}{\partial t} + \sum_K \left[\frac{\mathbf{P}_K}{M} \frac{\partial f^N}{\partial \mathbf{R}_K} + (\mathbf{F}_K + \mathbf{F}_K^A) \frac{\partial f^N}{\partial \mathbf{P}_K} \right]$$
$$= \sum_K \frac{\partial}{\partial \mathbf{P}_K} \left[- \int_{\Sigma_K} ds\, \mathbf{n} \cdot \left(p_2(\mathbf{x}, t) \mathbf{U} + \langle \mathbf{T}(\mathbf{x}) \Delta \rangle'_0 \right) f^N + kT \sum_L \zeta_{KL} \frac{\partial f^N}{\partial \mathbf{P}_L} \right] \tag{2.14}$$

and the friction coefficient (2.6) in the form

$$\zeta_{KL} = \beta \int_{-\infty}^{0} dt\, \langle \mathcal{F}_K \mathcal{F}_L(t) \rangle'_q \, . \tag{2.15}$$

The main distinction from the case of a single large particle discussed in the preceding chapter is that the velocity field $\mathbf{v}(\mathbf{x})$ here must to satisfy N boundary conditions

$$\mathbf{v}(\mathbf{x})|_{\Sigma_K} = \mathbf{P}_K / M \tag{2.16}$$

The surface integral in the right–hand side of equation (2.14) presents the mean force acting on one of the particles immersed in the viscous liquid. Its calculation reduces to a pure hydrodynamic problem of determining the velocity field satisfying the boundary condition (2.16) and subsequent integration over the surface, as in the preceding chapter.

In the general case of a high concentration of large particles, such solution is a cumbersome integral expression representative of the nonlocal character of hydrodynamic interparticle interactions in the system. And, only in the special case of a sufficiently rarefied dispersed system, when mean interparticle distance is much larger then the particle radius, can the mean force acting on a particle be represented [68] in the form

$$\langle \mathcal{F}_k \rangle = -\xi_{KK}\left(\frac{\mathbf{P}_K}{M} - \mathbf{v}_{1K}\right) - \sum_{L\neq K} \xi_{KL}\left(\frac{\mathbf{P}_L}{M} - \mathbf{v}_{1L}\right) \tag{2.17}$$

where $\mathbf{v}_{1K} \equiv \mathbf{v}_1(\mathbf{R}_K, t)$ and ζ_{KL} are the coefficients of hydrodynamic interaction between the Kth and Lth particles. There are many works devoted to calculations of these coefficients in the form of power series expansions in a/R_{KL}. In these, it is found that every coefficient depends on the coordinates of all particles of the system. Divergent integrals appear in averaging over these coordinates, which is a consequence of the long–range character ($\sim 1/R_{KL}$) of the hydrodynamic interactions. This representing the coefficients ζ_{KL} is not useful *per se*. In the hydrodynamics of suspensions, methods of renormalization for such an expansion have been developed [69, 70]. As these methods are special and different for particular forms of dispersed system motion, we will not dwell upon them here, and we will refer later to two of them when discussing the Brownian particle diffusion. Turning now to the friction coefficient (2.15), multiplying it by the tensor $(P_K^\alpha/M - v_{1K}^\alpha)(P_L^\beta/M - v_{1L}^\beta)$ and summing indices α, β and K, L using Gauss' theorem, we obtain

$$\sum_{K,L}\left(\frac{P_K^\alpha}{M} - v_{1K}^\alpha\right)\left(\frac{P_L^\beta}{M} - v_{1L}^\beta\right)\zeta_{KL}^{\alpha\beta} = \beta\int d\mathbf{x}\int d\mathbf{x}'\int_{-\infty}^{0} dt\frac{\partial}{\partial x_\mu}v_2^\alpha(\mathbf{x})$$
$$\times\frac{\partial}{\partial x'_\nu}v_2^\beta(\mathbf{x}')\left\langle T_{\alpha\mu}(\mathbf{x})\left(T_{\beta\nu}(\mathbf{x}',t) - \langle T_{\beta\nu}\rangle_0\right)\right\rangle_0 \tag{2.18}$$

In local approximation, the right–hand side of this equation can be rewritten in the form

$$\begin{aligned}
\int d\mathbf{x}\, \frac{\partial v_2^\alpha}{\partial x_\mu} \frac{\partial v_2^\beta}{\partial x_\nu}\, \beta \int_{-\infty}^{0} dt \int d\mathbf{x}' \left\langle T_{\alpha\mu}(\mathbf{x}) \left(T_{\beta\nu}(\mathbf{x}', t) - \langle T_{\beta\nu}\rangle_0\right)\right\rangle_0 \\
= \eta \int d\mathbf{x}\, \frac{\partial v_2^\alpha}{\partial x_\mu} \frac{\partial v_2^\beta}{\partial x_\nu} \left[\delta_{\alpha\beta}\delta_{\mu\nu} + \delta_{\alpha\nu}\delta_{\mu\beta} - \frac{2}{3}\delta_{\alpha\mu}\delta_{\beta\nu}\right] \\
= \eta \int d\mathbf{x} \left[\frac{\partial}{\partial x_\alpha} v_2^\beta \frac{\partial v_2^\alpha}{\partial x_\beta} + \frac{\partial}{\partial \mathbf{x}_\beta} v_2^\alpha \frac{\partial v_2^\alpha}{\partial x_\beta} - v_2^\alpha \frac{\partial^2}{\partial x_\beta^2} v_2^\alpha\right]
\end{aligned} \tag{2.19}$$

where the Kubo formula for the viscosity coefficient (5.5.5) and the incompressibility condition $\partial v_2^\alpha / \partial x_\alpha = 0$ are used. Employing the Stokes equation $\partial \sigma_{\alpha\beta}/\partial x_\alpha = 0$ and equations (2.16) and (2.17), we get

$$\begin{aligned}
\sum_{KL} \left(\frac{P_K^\alpha}{M} - v_{1K}^\alpha\right)\left(\frac{P_L^\beta}{M} - v_{1L}^\beta\right)\zeta_{KL}^{\alpha\beta} = -\sum_K \left(\frac{P_K^\alpha}{M} - v_{1K}^\alpha\right)\int_{\Sigma_K} ds\, n_\beta \sigma_{\alpha\beta} \\
= \sum_{KL} \left(\frac{P_K^\alpha}{M} - v_{1K}^\alpha\right)\left(\frac{P_L^\beta}{M} - v_{1L}^\beta\right)\xi_{KL}^{\alpha\beta}
\end{aligned} \tag{2.20}$$

As a consequence of mutual independence of the momenta P_K^α we find that $\zeta_{KL}^{\alpha\beta} = \xi_{KL}^{\alpha\beta}$. This fluctuation–dissipation relation removes the problem of calculating a correlation function: the right–hand side of equation (2.15) can be replaced by the much simpler problem of calculating the coefficients $\xi_{KL}^{\alpha\beta}$ from equation (2.17) for the mean force.

Substituting equation (2.17) and the derived fluctuation–dissipation relation into equation (2.14), we get the final kinetic equation for the N–particle distribution function in the form

$$\begin{aligned}
\frac{\partial f^N}{\partial t} + \sum_K \left[\frac{\mathbf{P}_K}{M}\frac{\partial f^N}{\partial \mathbf{R}_K} + (\mathbf{F}_K + \mathbf{F}_K^A)\frac{\partial f^N}{\partial \mathbf{P}_K}\right] \\
= \sum_{KL} \frac{\partial}{\partial \mathbf{P}_K} \xi_{KL} \left[\left(\frac{\mathbf{P}_L}{M} - \mathbf{v}_{1L}\right) f^N + kT \frac{\partial f^N}{\partial \mathbf{P}_L}\right]
\end{aligned} \tag{2.21}$$

The resultant equation is the many–particle generalization of the Fokker–Planck equation (5.5.18). In the special case of no dissipative interactions in the system (for example, the liquid viscosity is negligible), the right–hand side of this equation becomes zero, and it takes the form of the Liouville equation for the system of particles in the outer field of force $\mathbf{F}^A$, with the interparticle interaction force $\mathbf{F}_K$. The diagonal terms $(K = L)$ in the right–hand side of equation (2.21) describe the dissipative interaction of the particles with the surroundings.

The coefficient ξ_{KK} differs from the Stokes coefficient, as it characterizes the drag in the disperse media rather than in a pure liquid. The summation with the nondiagonal coefficient ξ_{KL} $(K \neq L)$ describes the hydrodynamic interparticle interactions induced by liquid disturbances from the particle motion. Notice that the right–hand side of the equation (2.21) vanishes if the function f^N is the local Maxwellian.

A similar equation for the system of suspended point heavy particles with potential (rather than hydrodynamic) interaction between particles and liquid was derived in [71, 72, 73] a thermodynamically equilibrium liquid. These works can be regarded as a development of the approach discussed in Section 5.8–5.10 of the preceding chapter. The coefficients derived there are expressed in terms of short–range potential interaction forces between the particles and molecules of the liquid. It is clear that as a consequence of their short–range character, the cross coefficients $\tilde{\xi}_{KL}$ $(K \neq L)$ are to vanish in real suspensions that do not obviously correspond to the physical properties of these systems. So in the following derivations (see, as an example, [72, 74, 75]), the coefficients $\tilde{\xi}_{KL}$ were considered to be hydrodynamic nature, and thus at variance with the formal prerequisites of the works cited [71, 72, 73].

Naturally, the hydrodynamic coefficients are independent of the material of which the Brownian particles are consisted and, therefore, of the potential of their interaction with the liquid molecules. Besides, in assuming a hydrodynamic interaction, one departs from the assumption of a thermodynamically equilibrium state of the liquid.

Both of these statements indicate that the approach developed in [71, 72, 73] is incompatible with the real nature of the hydrodynamic interparticle interactions in suspensions.

Here we have developed a more consistent derivation of the N–particle Fokker–Planck equation for a system of large particles in a viscous liquid; this is the development of the approach to the Brownian motion theory discussed in Section 5.2–5.7 of the preceding Chapter 5.

In Sec. 5.7 of Chapter 5 the limitation from below of the Brownian particle material density was introduced to provide time–scale separation between fluid and Brownian particle velocities. For many–particle systems such limitation become stronger, as in the case of unsufficiently massive suspended particles, the system cannot be described by local Markov equations [76, 77].

6.3 The Smoluchowski Equation

Modern phenomenological theories of disperse systems usually take, as their starting point, the Smoluchowski equation. Thus; it is interesting to derive this equation starting from the Fokker–Planck kinetic equation (2.21).

We integrate equation (2.21) over all particle momenta, $\{\mathbf{P}_K\}$. Then for N–particle density

$$C_N(\{\mathbf{R}_K\},t) = \int d\mathbf{P}_1 \ldots d\mathbf{P}_N f^N(\{\mathbf{P}_K,\mathbf{R}_K\},t) \tag{3.1}$$

we get the first moment equation in the form

$$\frac{\partial}{\partial t}C^N + \sum_K \frac{\partial}{\partial \mathbf{R}_K} C^N \mathbf{v}_{1K} + \sum_K \frac{\partial}{\partial \mathbf{R}_K} C^N (\mathbf{V}_K - \mathbf{v}_{1K}) = 0 \tag{3.2}$$

where

$$C_N \mathbf{V}_K = \langle \mathbf{P}_K / M \rangle \ , \quad \langle A \rangle = \int d\mathbf{P}_1 d\mathbf{P}_2 \ldots d\mathbf{P}_N \, A$$

The second term in this equation describes the convective transport of the particles in N-dimensional space, and the third term describes the diffusion transport. As the mean velocity $\mathbf{V}_K$ of the Brownian particles is unknown, equation (3.2) is not closed. To close it we use a second moment equation that is derived by way of multiplying equation (2.21) by $\mathbf{P}_K - M\mathbf{v}_{1K}$, and subsequent integrating over all momenta $\mathbf{P}_K$. As a result we get

$$\begin{aligned} M\frac{\partial}{\partial t}C^N(\mathbf{V}_K - \mathbf{v}_{1K}) &+ \sum_L \frac{\partial}{\partial \mathbf{R}_L} \left\langle \frac{(\mathbf{P}_K - M\mathbf{v}_{1K})\mathbf{P}_L}{M} \right\rangle \\ &+ \sum_L (\mathbf{F}_L + \mathbf{F}_L^A) \left\langle (\mathbf{P}_K - M\mathbf{v}_{1K}) \frac{\partial}{\partial \mathbf{P}_L} \right\rangle \\ &= \sum_{LM} \left\langle (\mathbf{P}_K - M\mathbf{v}_{1K}) \frac{\partial}{\partial \mathbf{P}_M} \xi_{ML} \left[\left(\frac{\mathbf{P}_L}{M} - \mathbf{v}_{1L} \right) + kT \frac{\partial}{\partial \mathbf{P}_L} \right] \right\rangle \end{aligned} \tag{3.3}$$

If we confine ourselves here to a stationary approximation (in diffusion velocity), not accounting for the acceleration $\frac{\partial}{\partial t}(\mathbf{V}_K - \mathbf{v}_K)$, then the first term in the left–hand side of equation (3.3) and the nondiagonal part of the second term with $L \neq K$ can be rewritten, considering equation (5.2) in the form

$$M(\mathbf{V}_K - \mathbf{v}_{1K}) \left[\frac{\partial}{\partial t} C^N + \sum_{L(\neq K)} \frac{\partial}{\partial \mathbf{R}_L} C^N \mathbf{V}_L \right] = -M \frac{\partial}{\partial \mathbf{R}_K} (\mathbf{V}_K - \mathbf{v}_{1K}) \mathbf{V}_K C^N \tag{3.4}$$

Then, the first two terms in the left–hand side of equation (3.3) take the form

$$\frac{\partial}{\partial \mathbf{R}_K}\left[\left\langle \frac{(\mathbf{P}_K - M\mathbf{v}_{1K})(\mathbf{P}_K - M\mathbf{v}_{1K})}{M}\right\rangle - M(\mathbf{V}_K - \mathbf{v}_{1K})(\mathbf{V}_K - \mathbf{v}_{1K})C^N\right]$$
$$= \frac{\partial}{\partial \mathbf{R}_K}\left\langle \frac{(\mathbf{P}_K - M\mathbf{v}_{1K})(\mathbf{P}_K - M\mathbf{v}_{1K})}{M}\right\rangle = \frac{\partial}{\partial \mathbf{R}_K} kT(\mathbf{R}_K, t)C^N \tag{3.5}$$

where $T(\mathbf{R}, t)$ is the temperature of the translational thermal motion of suspended particles.

The remaining terms of equation (3.3) can be represented as

$$\sum_L (\mathbf{F}_L + \mathbf{F}_L^A)\left\langle \mathbf{P}_k - M\mathbf{v}_{1K}\frac{\partial}{\partial \mathbf{P}_L}\right\rangle = -(\mathbf{F}_k + \mathbf{F}_K^A)C^N \tag{3.6}$$

$$\sum_{ML}\left\langle (\mathbf{P}_K - M\mathbf{v}_{1K})\frac{\partial}{\partial \mathbf{P}_M}\xi_{ML}\left[\left(\frac{\mathbf{P}_L}{M} - \mathbf{v}_{1L}\right) + kT\frac{\partial}{\partial \mathbf{P}_L}\right]\right\rangle$$
$$= -\sum_L \xi_{KL}(\mathbf{V}_L - \mathbf{v}_{1L})C^N \tag{3.7}$$

Substituting equations (3.5) through (3.7) into equation (3.3) we get

$$kT\frac{\partial}{\partial \mathbf{R}_K}C^N + C^N kT\frac{\partial}{\partial \mathbf{R}_K}\ln T - (\mathbf{F}_K + \mathbf{F}_K^A)C^N$$
$$= -\sum_L \xi_{KL}(\mathbf{V}_L - \mathbf{v}_{1L})C^N \tag{3.8}$$

The right–hand side of this equation can be regarded as a product of the matrix $\vec{\vec{\xi}} = \|\xi_{KL}\|$ by the $3N$–dimensional vector $\vec{\mathbf{V}}$ with the coordinates $\mathbf{V}_K$. Hereafter, the single and double arrows are used to denote vectors and matrices with indexes corresponding to particle numbers $K, L = 1, 2, \ldots, N$. Assuming an existence of the inverse matrix $\vec{\vec{\xi}}^{-1}$, such that $\vec{\vec{\xi}}^{-1}\cdot\vec{\vec{\xi}} = \vec{\vec{\mathbf{U}}}$ ($\vec{\vec{\mathbf{U}}}$ — is the unit tensor $\mathbf{U}_{KL} = \delta_{KL}$), we can obtain the equation for the diffusion flux from equation (3.8)

$$\mathbf{J}_K \equiv C^N(\mathbf{V}_K - \mathbf{v}_{1K})$$
$$= -kT\xi_{KL}^{-1}\frac{\partial}{\partial \mathbf{R}_L}C^N - \xi_{KL}^{-1}kTC^N\frac{\partial}{\partial \mathbf{R}_L}\ln T + C^N\xi_{KL}^{-1}(\mathbf{F}_L + \mathbf{F}_L^A) \tag{3.9}$$

or, in the $3N$–vector notation

$$\vec{\mathbf{J}} = -\vec{\vec{\mathbf{D}}} \cdot \vec{\nabla} C^N - \vec{\vec{\mathbf{D}}}_T \cdot \frac{\vec{\nabla} T}{T} + \vec{\vec{\mathbf{D}}} \cdot \frac{\vec{\mathbf{F}} + \vec{\mathbf{F}}^A}{kT} C^N \tag{3.10}$$

where

$$\vec{\vec{\mathbf{D}}} = kT \vec{\vec{\xi}}^{-1} \tag{3.11}$$

is the generalized Einstein relation for the diffusion coefficient and

$$\vec{\vec{\mathbf{D}}}_T = kT \vec{\vec{\xi}}^{-1} C^N = \vec{\vec{\mathbf{D}}} C^N \tag{3.12}$$

is the analogous relation for the thermodiffusion coefficient.

Substituting equation (3.10) into equation (3.2) we get the generalized Smoluchowski equation in the final form

$$\frac{\partial}{\partial t} C^N + \vec{\nabla} \cdot C^N \vec{\mathbf{v}}_1 = \vec{\nabla} \cdot \vec{\vec{\mathbf{D}}} \cdot \left(\vec{\nabla} + \frac{\vec{\nabla} T}{T} - \frac{\vec{\mathbf{F}} + \vec{\mathbf{F}}^A}{kT} C^N \right) \tag{3.13}$$

This equation contains the convective term in the left–hand side and the thermodiffusion term and the Archimedean force on the right–hand side as an adjunct to the well–known form of this equation found extensively applied in disperse system theories. The additional terms allow for a nonequilibrium state of the liquid in which the particles are suspended.

The Smoluchowski equation (3.13) is unsuitable as such for use in applications because of its extreme complexity, namely, it describes the evolution of the N–particle coordinate distribution function in the $3N$-dimensional space. Its main value is that it can be used as a starting point for the theoretical studies of particular diffusion processes. In this respect, the generalized Smoluchowski equation takes a place in the diffusion processes theory similar to that of the Liouville equation in the kinetic theory.

Now we apply this equation to the Brownian diffusion of particles with the hydrodynamic interparticle interaction in nonisothermal viscous liquid flow.

6.4 Brownian Diffusion of Suspended Particles

To accomplish the derivation of the one–particle diffusion equation we introduce the S–particle coordinate distribution function in the form

$$C_S(\mathbf{R}_1, \ldots, \mathbf{R}_S, t) = \Omega^S \int d\mathbf{R}_{S+1} \ldots d\mathbf{R}_N \, C^N \tag{4.1}$$

where Ω is the system volume. The multiplier Ω^S ensures that the S–particle functions are undimensional. Their normalization conditions are

$$\frac{1}{\Omega^S}\int d\mathbf{R}_1\ldots d\mathbf{R}_S\, C_S(\mathbf{R}_1,\ldots,\mathbf{R}_S) = 1 \tag{4.2}$$

Whence it follows, in particular, that $C_1(\mathbf{R}_1, t)$ is identically equal to one in a space-homogeneous system.

Integrating equation (3.13) over all $\{\mathbf{R}_K\}$ but $\mathbf{R}_1$ we get

$$\begin{aligned}
&\frac{\partial}{\partial t}C_1(\mathbf{R}_1,t) + \nabla_1\cdot C_1\mathbf{v}_1\\
&= \Omega\nabla_1\cdot\int d\mathbf{R}_2\ldots d\mathbf{R}_N\,\mathbf{D}_{11}\cdot\left(\nabla_1 + \frac{\nabla_1 T}{T} - \frac{\mathbf{F}_1+\mathbf{F}_1^A}{kT}\right)C^N\\
&\quad + \Omega\nabla_1\cdot\int d\mathbf{R}_2\ldots d\mathbf{R}_N\sum_{L=2}^{N}\mathbf{D}_{1L}\cdot\left(\nabla_L + \frac{\nabla_L T}{T_1} - \frac{\mathbf{F}_L+\mathbf{F}_L^A}{kT}\right)C^N
\end{aligned} \tag{4.3}$$

where $\nabla_K \equiv \partial/\partial\mathbf{R}_K$, $\mathbf{v}_1$, $\mathbf{F}_K$, $\mathbf{F}_K^A$ are 3–dimensional vectors, $\mathbf{D}_{11}$, $\mathbf{D}_{1L}$ are 3–dimensional tensors, and the dot is used to denote a scalar product in three–dimensional Cartesian space.

The tensor $\mathbf{D}_{11}$ depends on all suspended particle coordinates. The probability of an influence of any particle on the motion of the selected particle is proportional to the mean concentration of the particles $n_0 = N/\Omega$, and the probability of a third particle disturbing this influence is proportional to n_0^2. Thus, for sufficiently rarefied systems, one may use an approximation of pair hydrodynamic interactions, or to consider the tensor $\mathbf{D}_{11}$ to be dependent only on $\mathbf{R}_1$ and $\mathbf{R}_2$ and having the form

$$\mathbf{D}_{11} = D_0\mathbf{U} + \sum_{L=2}^{N}\mathbf{A}(\mathbf{R}_1,\mathbf{R}_L) \tag{4.4}$$

where $D_0 = kT/\zeta_S$ and $\mathbf{U}$ is the unit tensor and $\mathbf{A}(\mathbf{R}_1,\mathbf{R}_L)$ is a three–dimensional tensor depending on $\mathbf{R}_1$ and $\mathbf{R}_L$ only.

In this approximation equation (4.3) takes the form

$$
\begin{aligned}
&\frac{\partial C_1}{\partial t} + \nabla_1 C_1 \mathbf{v}_1 \\
&= D_0 \nabla_1 \left(\nabla_1 + \frac{\nabla_1 T}{T} - \beta F_1^A \right) C_1 \\
&+ D_0 n_0 \beta \nabla_1 \int d\mathbf{R}_2\, C_2 \nabla_1 \tilde{\Phi}_{12} \\
&+ n_0 \nabla_1 \int d\mathbf{R}_2\, \mathbf{A}(\mathbf{R}_1 \mathbf{R}_2) \left(\nabla_1 + \frac{\nabla_1 T}{T} - \beta \mathbf{F}_1^A \right) C_2(\mathbf{R}_1, \mathbf{R}_2, t) \\
&+ n_0^2 \beta \nabla_1 \int d\mathbf{R}_2 d\mathbf{R}_3\, \mathbf{A}(\mathbf{R}_1, \mathbf{R}_2) C_3(\mathbf{R}_1, \mathbf{R}_2, \mathbf{R}_3, t) \nabla_1 \tilde{\Phi}_{13} \qquad (4.5) \\
&+ n_0 \beta \nabla_1 \int d\mathbf{R}_2\, \mathbf{A}(\mathbf{R}_1, \mathbf{R}_2) C_2(\mathbf{R}_1, \mathbf{R}_2, t) \nabla_1 \tilde{\Phi}_{12} \\
&+ n_0 \nabla_1 \int d\mathbf{R}_2\, \mathbf{D}_{12} \left(\nabla_2 + \frac{\nabla_2 T}{T} - \beta F_2^A \right) C_2(\mathbf{R}_1, \mathbf{R}_2, t) \\
&+ n_0^2 \beta \nabla_1 \int d\mathbf{R}_2 d\mathbf{R}_3\, \mathbf{D}_{12} C_3(\mathbf{R}_1, \mathbf{R}_2, \mathbf{R}_3, t) \nabla_2 \tilde{\Phi}_{23} \\
&+ n_0 \beta \nabla_1 \int d\mathbf{R}_2\, \mathbf{D}_{12} C_2(\mathbf{R}_1, \mathbf{R}_2, t) \nabla_2 \tilde{\Phi}_{21}
\end{aligned}
$$

Here we have used the common assumtion about the symmetry of the function C_N with respect to particle permutations, and have substituted the explicit expressions of the hard–sphere forces $\mathbf{F}_1$ and $\mathbf{F}_2$, acting on the first and second particles from other particles:

$$
\mathbf{F}_1 = -\nabla_1 \sum_{L=2}^{N} \tilde{\Phi}_{1L}\,, \quad \mathbf{F}_2 = -\nabla_2 \tilde{\Phi}_{21} - \nabla_2 \sum_{L=3}^{N} \tilde{\Phi}_{2L} \qquad (4.6)
$$

As equation (4.5) is written in the approximation of pair hydrodynamic interactions, terms with n_0^2 are to be omitted. Then, in the right–hand side of equation (4.5) only the terms with C_1 and C_2 are conserved; this equation takes the form resembling the first equation of the BBGKY hierarchy in the kinetic gas theory.

To close this equation we represent the two-particle distribution function in the form

$$
C_2(\mathbf{R}_1, \mathbf{R}_2, t) = C_1(\mathbf{R}_1, t) C_1(\mathbf{R}_2, t) g_{12}(\mathbf{R}_1, \mathbf{R}_2, t) \qquad (4.7)
$$

where g_{12} is the radial distribution function. In a rarefied equilibrium system it takes the form $g_{12}^0 = \exp\{-\beta\tilde{\Phi}_{12}\}$. Since the function C_2 enters into equation (4.5) only through the terms containing the multipliers n_0, in the frame of considered approximation we may substitute equation (4.7) into equation (4.5), taking g_{12} in the form of g_{12}^0 when the approximation of low density for weakly nonequilibrium diffusion processes is considered.

Then, the second term in the linear approximation in the concentration gradient takes the form

$$\begin{aligned} &-n_0 D_0 C_1(\mathbf{R}_1,t)\int d\mathbf{R}_2\, C_1(\mathbf{R}_2,t)\beta\frac{\partial\tilde{\Phi}_{12}}{\partial\mathbf{R}_2}e^{-\beta\tilde{\Phi}_{12}} \\ &= -n_0 D_0 C_1(\mathbf{R}_1,t)\int d\mathbf{r}\, C_1(\mathbf{R}_1+\mathbf{r},t)\beta\frac{\partial\tilde{\Phi}(\mathbf{r})}{\partial\mathbf{r}}e^{-\beta\tilde{\Phi}(\mathbf{r})} \\ &= -4\pi n_0 D_0 C_1\nabla_1 C_1\int_{2a}^{\infty} dr\, r^2 e^{-\beta\tilde{\Phi}(\mathbf{r})} = 8D_0\varphi\nabla_1 C_1 \end{aligned} \tag{4.8}$$

where $\varphi = \frac{4}{3}\pi a^3 n_0$ is the volume concentration of the Brownian particles.

The third term in the right–hand side of equation (4.5) (in the approximation under consideration as a result of the action of the operator ∇_1 on g_{12}) is written as the sum of two terms:

$$\begin{aligned} &\int d\mathbf{R}_2\,\mathbf{A}\; g_{12}^0\left(\nabla_1+\frac{\nabla_1 T}{T}-\beta\mathbf{F}_1^A\right)C_1(R_1,t)C_1(R_2,t) \\ &\qquad -\beta\int d\mathbf{R}_2\,\mathbf{A}C_1(\mathbf{R}_1,t)C_1(\mathbf{R}_2,t)g_{12}^0\nabla_1\tilde{\Phi}_{12} \end{aligned} \tag{4.9}$$

Notice that the second term of this expression, when substituted into equation (4.5), will cancel with the fifth term. Remind that the fourth and seventh terms in the right–hand side of equation (4.5) are neglected since they are proportional to n_0^2. The sixth term can be written in the same manner as the third, in the form of two components:

$$\begin{aligned} &\int dR_2\;\mathbf{D}_{12}g_{12}^0\left(\nabla_2+\frac{\nabla_2 T}{T}-\beta\mathbf{F}_2^A\right)C_1(\mathbf{R}_1,t)C_1(\mathbf{R}_2,t) \\ &\qquad -\beta\int d\mathbf{R}_2\,\mathbf{D}_{12}C_1(\mathbf{R}_1,t)C_1(\mathbf{R}_2,t)g_{12}^0\nabla_2\tilde{\Phi}_{12} \end{aligned} \tag{4.10}$$

The second component is coincident (with the opposite sign) with the last, or eighth, term in the right–hand side of equation (3.3). Since the pair interaction tensor $\mathbf{D}_{12}$ differs from zero only at not too much great distances between the first and the second particles the first integrand in equation (4.10) may be rewritten in the linear approximation in gradients as

$$\left(\nabla_2+\frac{\nabla_2 T}{T}-\beta\mathbf{F}_2^A\right)C_1(\mathbf{R}_2,t)\simeq\left(\nabla_1+\frac{\nabla_1 T}{T}-\beta\mathbf{F}_1^A\right)C_1(\mathbf{R}_1,t) \tag{4.11}$$

Substituting equations (4.8) through (4.11) into equation (4.5) and multiplying all terms of the resultant equation by the particle number N, we get the equation for the convective diffusion of Brownian particles in an inhomogeneous fluid in the form

$$\frac{\partial n}{\partial t} + \nabla n \mathbf{v}_1 = \nabla D \left(\nabla + \frac{\nabla T}{T} + \frac{4}{3}\pi a_0^3 \frac{\nabla p_1}{kT} \right) n \tag{4.12}$$

where $n(\mathbf{r}, t) = NC_1(\mathbf{r}, t)$ is the particle number density, and

$$D = D_0 \left[1 + 8\varphi + n_0 \int d\mathbf{R}_2\, \mathbf{A}_{12} C_1(R_2, t) + n_0 \int d\mathbf{R}_2\, \mathbf{D}_{12} C_1(\mathbf{R}_2, t) \right] \tag{4.13}$$

is the diffusion coefficient. The term 8φ in equation (4.13) describes an increase of the diffusion flow, due to the pure geometric effect of excluding the volume, filled with other particles, from the whole volume of the system.

To determine two integrals in equation (4.13) it is necessary to solve the hydrodynamic problem of two bodies and to find the tensors $\mathbf{A}_{12}$ and $\mathbf{D}_{12}$. The central difficulty here is associated with the unsuitability of power series expansions in a/R_{12}, as they lead to divergent integrals in equation (4.13). An elegant solution to this problem has been found by Batchelor [78]. In a lowest approximation in particle concentration, he succeeded in regrouping separate terms describing hydrodynamic disturbances induced by two particles motions such that divergences cancelled each other. As a result, he derived

$$n_0 \int d\mathbf{R}_2\, \mathbf{A}_{12} C_1(\mathbf{R}_2, t) = -1.83\varphi \tag{4.14}$$

$$n_0 \int d\mathbf{R}_2\, \mathbf{D}_{12} C_1(\mathbf{R}_2, t) = -4.72\varphi \tag{4.15}$$

A negative value for the first integral reflects a retardation of the diffusion motion of the selected particle by the presence of another particle *per se*. The second is negative also, meaning that the diffusion motion of another particle retards the selected particle motion. Thus, equation (4.13) takes the final form

$$D = D_0(1 + 1.45\varphi) \tag{4.16}$$

Introducing the thermodiffusion and barodiffusion coefficients as

$$D_T = nD\,, \quad D_p = \varphi D/kT \tag{4.17}$$

we rewrite the diffusion equation (4.12) in the final form

$$\frac{\partial n}{\partial t} + \nabla n \mathbf{v}_1 = \nabla (D \nabla n + D_T \nabla T/T + D_p \nabla p_1) \tag{4.18}$$

The resultant expression for D coincides with the result of Batchelor's [78] phenomenological theory, founded on the use of the Einstein relations, modified to allow for a finite volume concentration of particles. It was this specific modification that led to the discussion [66, 67] of the Batchelor result. This approach make it possible to avoid introducing any additional relations and thus thus substantiates the choice of the Batchelor theory. The Einstein relation was also derived in the form of equation (3.11). Also, the diffusion and thermodiffusion coefficients in a suspension were obtained.

6.5 Convective Diffusion of Particles in Fixed Dispersed Layers

The last sections of this chapter are devoted to Brownian diffusion of particles suspended in liquid flow running through a stochastic system of fixed particles randomly distributed in space. The flow around each fixed particle generates a flow perturbation of the liquid velocity. As a consequence of a randomness of the array of these particles, the perturbations are also random. Superposed on thermal fluctuations of the liquid velocity, they result in a modification of the diffusion coefficients of the suspended Brownian particle, derived in Chapter 5, allowing for thermal fluctuations only.

In this section we derive the equation for a convective diffusion of a Brownian particle in the flow running through a grid of fixed particles [79].In the next sections, we will consider a stochastic equation for this system allowing for the random perturbation of the flow velocity. Using the Green function for this equation, the root–mean–square deviation of Brownian particles across the mean flow direction, with an effective, renormalized diffusion coefficient, is derived.

When deriving the convective diffusion equation we begin with the N–particle Fokker–Plancke equation (2.21) rather than the generalized Smoluchowski equation (3.13). This preference is because the analysis of many–particle hydrodynamic interactions in such systems can be more conveniently performed in terms of the drag coefficients ξ_{KL} rather than the diffusion coefficients D_{KL}.

It was found [70, 80, 81] in the hydrodynamic theory of a particle motion in a porous medium (formed by a rarefied system of fixed spheres of the volume concentration φ_2) that the main correction to the viscous drag force was of the order of $\sqrt{\varphi_2}$. This fact permits us to use the point particle hydrodynamic approximation as the error of such an approximation; that is, an error of the order of $O(\varphi_2)$ does not affect the accuracy of an expected result of the order of $O(\sqrt{\varphi_2})$.

Therefore, the terms with $\mathbf{F}_K$ and $\mathbf{F}_K^A$ may be ignored as being of the order of φ.

The subsequent discussion will be confined to the particular case of an isothermal liquid flow of a homogeneous inherent velocity $\mathbf{v}_1 = \mathbf{v}_0$.

In the system under consideration the first particle, of radius a_1, is suspended in the liquid flow, and all other particles, of radius a_2, are fixed at random coordinates $\mathbf{R}_2, \ldots, \mathbf{R}_N$.

A diffusion equation for the suspended particle is derived as the first moment equation of the Fokker–Planck equation (3.21) by the way of integration over coordinates of all fixed particles $\mathbf{R}_2, \ldots, \mathbf{R}_N$ and all momenta $\mathbf{P}_1, \ldots, \mathbf{P}_N$. Then we get

$$\frac{\partial}{\partial t}C + \frac{\partial}{\partial \mathbf{R}_1}C\mathbf{v}_0 + \frac{\partial}{\partial \mathbf{R}_1}C\mathbf{V}_d = 0 \tag{5.1}$$

where

$$C(\mathbf{R}_1, t) = \langle\langle 1 \rangle\rangle\ , \quad C(\mathbf{R}_1, t)\mathbf{V}_d(\mathbf{R}_1, t) = \left\langle\left\langle \frac{\mathbf{P}_1}{M} - \mathbf{v}_0 \right\rangle\right\rangle \tag{5.2}$$

and

$$\langle\langle A \rangle\rangle \equiv \int d\mathbf{R}_2 \ldots d\mathbf{R}_N d\mathbf{P}_1 \ldots d\mathbf{P}_N\, A f_N \tag{5.3}$$

The second and third terms in equation (5.1) describe convective and diffusion transfers of particles, respectively. Since the diffusion velocity $\mathbf{V}_d$ is unknown, equation (5.1) is not closed. We have discussed this problem in Section 6.3 when deriving the Smoluchowski equation. Here also we use the second moment equation in a stationary approximation, that is

$$\frac{\partial}{\partial \mathbf{R}_1}CkT = -C\mathbf{V}_d\overline{\xi_{11}} + C\sum_{L\neq 1}\overline{\xi_{1L}}\ v v_0 \tag{5.4}$$

where the translation temperature $T(\mathbf{R}_1, t)$ is introduced by the equation

$$\langle\langle \mathbf{P}_1\mathbf{P}_L \rangle\rangle = \delta_{1L}MC(\mathbf{R}_1, t)kT(\mathbf{R}_1, t)$$

and the other momenta are represented as

$$\left\langle\left\langle \mathbf{P}_1\frac{\partial}{\partial \mathbf{P}_K}\xi_{KK}\left(\frac{\mathbf{P}_K}{M} - \mathbf{v}_0 + kT\frac{\partial}{\partial \mathbf{P}_K}\right)\right\rangle\right\rangle = -\delta_{1K}C(\mathbf{R}_1, t)\mathbf{V}_d\overline{\xi_{11}} \tag{5.5}$$

$$\left\langle\left\langle \mathbf{P}_1\frac{\partial}{\partial \mathbf{P}_K}\sum_{L(\neq K)}\xi_{KL}\left(\frac{\mathbf{P}_L}{M} - \mathbf{v}_0 + kT\frac{\partial}{\partial \mathbf{P}_L}\right)\right\rangle\right\rangle = -\delta_{1K}\mathbf{v}_0C(\mathbf{R}_1, t)\sum_{L(\neq 1)}\overline{\xi_{1L}} \tag{5.6}$$

Here it is taken into account that $\mathbf{V}_K \equiv \mathbf{P}_K/M = \delta_{1K}\mathbf{P}_1/M$, and the bar denotes averaging over the coordinates of fixed particles.

From equation (5.4) we get the diffusion flow in the form

$$\mathbf{J} \equiv CV_d = -kT\overline{\xi_{11}^{-1}}\nabla C + \overline{\xi_{11}^{-1}} \sum_{K(\neq 1)} \overline{\xi_{1K}}\mathbf{v}_0 C \tag{5.7}$$

Substituting (5.7) into the first moment equation (5.1), we get the convection diffusion equation

$$\frac{\partial}{\partial t}C + (\mathbf{v}_0 + \overline{\mathbf{v}})\nabla C - D\nabla^2 C = 0 \tag{5.8}$$

where

$$D = \overline{\xi^{-1}}kT \tag{5.9}$$

is the diffusion coefficient and

$$\overline{\mathbf{v}} = \overline{\xi_{11}^{-1}} \sum_{K} \overline{\xi_{1K}}\mathbf{v}_0 \tag{5.10}$$

is the mean perturbation of the flow velocity due to flow passage through the random configuration of fixed particles ($K = 2, 3, \ldots, N$).

6.6 Stochastization of the Diffusion Equation

The subsequent analysis of equation (5.8) closely resembles the Bedeaux and Mazur [82] approach. They have investigated the influence of thermal hydrodynamic fluctuation of a fluid velocity field on the particle diffusion in a homogeneous fluid, and found that the diffusion coefficient of the large particle is entirely determined by these fluctuations. In our approach, these fluctuations have been taken into account at the molecular level of description when deriving the Fokker–Planck equation. Here stochastic perturbations of the convective velocity of the flow around the random fixed sphere configuration are considered in addition to the thermal fluctuations. So as a "bare" diffusion coefficient we use in equation (5.7) the expression (5.9) obtained in terms of thermal fluctuations of the fluid stress tensor, whereas in [82] a "bare" diffusion coefficient has been taken to be zero.

As evidenced by the foregoing, the derivation of the convective diffusion equation from the Fokker–Planck equation involves the averaging of convective flow perturbation over random coordinates of the fixed spheres. Taking into consideration fluctuations of this perturbation causes an additional renormalization of the diffusion coefficient.

Instead of the diffusion equation (5.8) we consider the relevant stochastic Langevin equation where the stochastic value $\tilde{\mathbf{v}} = \xi_{11}^{-1} \sum_K \xi_{1K} \mathbf{v}_0$ substitutes the mean perturbation $\overline{\mathbf{v}}$ of the flow velocity, equation (5.10). Introducing the δ–shaped source term into the right–hand side of this equation, we write it for the Green function, as the propagator, describing the possibility of finding the particle at some point $\mathbf{r}$ at time t, under the condition that it was at $\mathbf{r}_0$ at the initial time t_0. Thus, we get

$$\left(\frac{\partial}{\partial t} - D\nabla^2 + (\mathbf{v}_0 + \tilde{\mathbf{v}}(\mathbf{r})\nabla)\right) G(\mathbf{r}, \mathbf{r}_0, t, t_0) = \delta(\mathbf{r} - \mathbf{r}_0)\delta(t - t_0) \quad (6.1)$$

with the boundary conditions $G(\mathbf{r}, \mathbf{r}_0, t, t_0) = 0$ when $t = t_0$ and $(\nabla G \mathbf{n}) = 0$ at the surfaces of every particle, n is the normal to the surface. According to the propagator definition we have

$$\langle(\mathbf{r} - \mathbf{r}_0)^n\rangle_d = \int d\mathbf{r}\, (\mathbf{r} - \mathbf{r}_0)^n G(\mathbf{r}, \mathbf{r}_0, t, t_0) \quad (6.2)$$

whence in view of equation (6.1) in the point particle approximation, disregarding surface integrals, we get

$$\frac{\partial}{\partial t} \langle(\mathbf{r} - \mathbf{r}_0)\rangle_d = \mathbf{v}_0 + \langle\tilde{\mathbf{v}}(\mathbf{r})\rangle \quad (6.3)$$

$$\frac{\partial}{\partial t} \langle(r_\alpha - r_{0\alpha})^2\rangle_d = 2D\Theta(t) + 2\langle(r_\alpha - r_{0\alpha})(v_{0\alpha} + \tilde{v}_\alpha(\mathbf{r}))\rangle \quad (6.4)$$

To calculate the right–hand sides of these equations it is necessary to find the explicit form of the propagator $G(\mathbf{r}, \mathbf{r}_0, t, t_0)$. With this in mind we write the Fourier transform of equation (6.1) in the form

$$(-i\omega - D\nabla^2 + \mathbf{v}_0 \cdot \nabla)G(\mathbf{r}, \mathbf{r}_0, \omega) = \delta(\mathbf{r} - \mathbf{r}_0) - \tilde{\mathbf{v}}(\mathbf{r}) \cdot \nabla G(\mathbf{r}, \mathbf{r}_0, \omega) \quad (6.5)$$

whence we find the integral equation

$$G(\mathbf{r}, \mathbf{r}_0, \omega) = G_0(\mathbf{r}, \mathbf{r}_0, \omega) - \int d\mathbf{r}'\, G_0(\mathbf{r}, \mathbf{r}', \omega)\tilde{\mathbf{v}}(\mathbf{r}') \cdot \nabla_{r'} G(\mathbf{r}', \mathbf{r}_0, \omega) \quad (6.6)$$

where $G_0(\mathbf{r}, \mathbf{r}_0, \omega)$ is the Green function of the unperturbed equation

$$(-i\omega - D\nabla^2 + \mathbf{v}_0 \cdot \nabla)G_0(\mathbf{r}, \mathbf{r}_0, \omega) = \delta(\mathbf{r} - \mathbf{r}_0) \,, \quad (6.7)$$

By iteration we find the solution to the integral equation (6.6) in the form of the power series in the velocity perturbation $\tilde{\mathbf{v}}$:

$$G(\mathbf{r}, \mathbf{r}_0, \omega) = G_0(\mathbf{r}, \mathbf{r}_0, \omega) - \int d\mathbf{r}'\, G_0(\mathbf{r}, \mathbf{r}', \omega)\tilde{\mathbf{v}} \cdot \nabla_{r'} G_0(\mathbf{r}', \mathbf{r}_0, \omega) \quad (6.8)$$

Here we confine our solution to the linear approximation in $\tilde{\mathbf{v}}(\mathbf{r})$. Substituting the solution (6.8) into equation (6.4) the mean–squared displacement across the flow $\mathbf{v}_0$ is found in the form

$$-i\omega\left\langle (x-x_0)^2\right\rangle_d = \frac{2Di}{\omega} + 2\int d\mathbf{r}\,(x-x_0)\tilde{v}_x(\mathbf{r})G_0(\mathbf{r}-\mathbf{r}_0,\omega)$$
$$-2\int d\mathbf{r}d\mathbf{r}'\,(x-x_0)G_0(\mathbf{r}-\mathbf{r}_0',\omega)\tilde{v}_x(\mathbf{r})\tilde{\mathbf{v}}(\mathbf{r}')\nabla G_0(\mathbf{r}'-\mathbf{r}_0,\omega) \tag{6.9}$$

Introducing into equation (6.9) the new variables $\rho = \mathbf{r}-\mathbf{r}'$, $\rho' = \mathbf{r}'-\mathbf{r}_0$, and integrating with respect to each of them over the total volume, introducing an error of the order of φ, we get

$$-i\omega\left\langle (x-x_0)^2\right\rangle_d = \frac{2Di}{\omega} - 2\int d\rho d\rho'\,\rho_x G_0(\rho-\rho',\omega)\Gamma_{x\beta}(\rho-\rho')\frac{\partial}{\partial\rho'_\beta}G_0(\rho',\omega) \tag{6.10}$$

where the new expression $\Gamma_{x\beta}(\mathbf{r}-\mathbf{r}') = \tilde{v}_x(\mathbf{r})\tilde{v}_\beta(\mathbf{r}')$ is introduced. The integral term in equation (6.10) is the convolution, so it is convenient to perform the Fourier space transform, then

$$-i\omega\overline{\langle (x-x_0)^2\rangle_d} = \frac{2Di}{\omega} + \frac{2i}{(2\pi)^3\omega}\int d\mathbf{k}\,\Gamma_{xx}(-\mathbf{k})G_0(\mathbf{k},\omega) \tag{6.11}$$

where the relations

$$\rho_x(\mathbf{k}) = -2\pi\frac{\partial}{\partial k_x}\delta(\mathbf{k})\,,\quad G_0(\mathbf{k},\omega) = \frac{1}{-i\omega + Dk^2 + i\mathbf{v}_0\mathbf{k}} \tag{6.12}$$

are used.

So it can be seen from equation (6.11) that in the limit $\omega \to 0$ the effective, renormalized coefficient of transverse diffusion is of the form

$$D_\perp = D + \Delta D_\perp = D + \frac{1}{(2\pi)^3}\int d\mathbf{k}\,\Gamma_{xx}(-\mathbf{k})G_0(\mathbf{k},0) \tag{6.13}$$

Explicit forms of all values in the integrand and the final form of this equation will be found in the next section.

6.7 Renormalized Diffusion Coefficient

We consider a particle moving with velocity $\mathbf{V}_K = \delta_{1K}\mathbf{V}_1$ in liquid with intrinsic velocity $\mathbf{v}_0+\tilde{\mathbf{v}}(\mathbf{r})$, that is, the velocity in the absence of the particle. This particle is subject to drag, (2.17) that can be also represented as

$$\langle\mathcal{F}_K\rangle = -\zeta_K(\mathbf{V}_K - \mathbf{v}_0 - \tilde{\mathbf{v}}_K) \tag{7.1}$$

where $\zeta_k = 6\pi\eta a_K$ is the Stokes coefficient for particle of radius $a_K = \delta_{1K}a_1 + (1-\delta_{1K})a_2$ and $\tilde{\mathbf{v}}_K \equiv \tilde{\mathbf{v}}(\mathbf{R}_K)$. This particle motion in turn generates a perturbation of the liquid flow that can be represented in the point particle approximation as

$$\Delta\mathbf{v}(\mathbf{r}) = \mathbf{T}(\mathbf{r}-\mathbf{R}_K)\langle -\mathcal{F}_k\rangle = \zeta_K \mathbf{T}(\mathbf{r}-\mathbf{R}_K)(\mathbf{V}_K - \mathbf{v}_0 - \tilde{\mathbf{v}}_K) \tag{7.2}$$

where $\mathbf{T}(\mathbf{r}) = (8\pi\eta r)^{-1}(\mathbf{U} + \mathbf{rr}/r^2)$ is the Oseen tensor; that is, the Green function of the Stokes equation, $\mathbf{U}$ is the unit tensor. Denoting for brevity $\mathbf{T}_{KL} = \mathbf{T}(\mathbf{R}_K - \mathbf{R}_L)$, $\mathbf{T}_{KK} = 0$, we write $\tilde{\mathbf{v}}_K$ in the form

$$\tilde{\mathbf{v}}_K = -\sum_L \zeta_L \mathbf{T}_{KL}(\mathbf{v}_0 - \mathbf{V}_L + \tilde{\mathbf{v}}_L) \tag{7.3}$$

whence by iteration we find

$$\begin{aligned}\tilde{\mathbf{v}}_K = &-\sum_L \zeta_L \mathbf{T}_{KL}(\mathbf{v}_0 - \mathbf{V}_L) + \sum_{LM} \zeta_L\zeta_M \mathbf{T}_{KM}\mathbf{T}_{ML}(\mathbf{v}_0 - \mathbf{V}_L) \\ &- \sum_{LMQ} \zeta_L\zeta_M\zeta_Q \mathbf{T}_{KQ}\mathbf{T}_{QM}\mathbf{T}_{ML}(\mathbf{v}_0 - \mathbf{V}_L) + \ldots\end{aligned} \tag{7.4}$$

From equations (2.17) and (7.1) for the mean force acting on the particle we have

$$\xi_{KL}(\mathbf{V}_L - \mathbf{v}_0) = \zeta_K(\mathbf{V}_K - \mathbf{v}_0 - \tilde{\mathbf{v}}_K) \tag{7.5}$$

Substituting here the expansion (7.4) of $\tilde{\mathbf{v}}_K$ and equating the coefficients attached to $\mathbf{V}_K - \mathbf{v}_0$ we get

$$\begin{aligned}\bar{\xi}_{KK} = \zeta_K &+ \zeta_K^2 \sum_L \zeta_L \overline{\mathbf{T}_{KL}\mathbf{T}_{LM}} - \zeta_K^2 \sum_{LM} \zeta_L\zeta_M \overline{\mathbf{T}_{KL}\mathbf{T}_{LM}\mathbf{T}_{MK}} \\ &+ \zeta_K^2 \sum_{LMQ} \zeta_L\zeta_M\zeta_Q \overline{\mathbf{T}_{KL}\mathbf{T}_{LM}\mathbf{T}_{MQ}\mathbf{T}_{QK}} - \ldots\end{aligned} \tag{7.6}$$

where the horizontal bar denotes the averaging over all particle coordinates, except the Kth one, with a homogeneous distribution function.

As all $\mathbf{T}_{KL}$ differ from zero only if $K \neq L$, all indexes are different in the terms of the second and third orders in $\mathbf{T}$. It is possible to have coincident indexes in the terms of higher orders. As an example, it may be that $L = Q$ and (or) $M = K$ in the term of the fourth order. At first we shall consider the expansion (7.6), assuming that all indexes are different. Then this expansion can be represented as the sum of the loop graphs

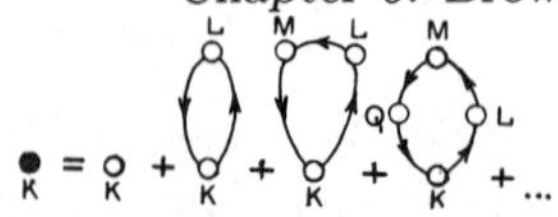

(7.7)

where the vertex • corresponds to the coefficient $\overline{\xi_{KK}}$ and the vertex ∘ corresponds to the Stokes coefficient ζ_K, and the directed lines connecting vertices correspond to relevant tensors $\mathbf{T}_{KL}$. The summation is implied over all indexes but K. The first term of this expansion (of the zeroth order in $\mathbf{T}$) describes the particle drag equal to ζ_K upon its moving through the pure liquid. The next term of the first order describes the contribution to the drag force from the process of reflection of the perturbation generated by the Kth particle from the Lth particle. The term of nth order describes the propagation of such a perturbation along the loop formed by n particles.

We are interested in the drag coefficient of the mobile Brownian particle ($K = 1$). Averaging all terms of the series (7.7) over fixed particles coordinates with the homogeneous distribution function, we get

$$\begin{aligned}\overline{\xi_{11}} = \zeta_1 + \zeta_1^2\zeta_2 n_2 \int d\mathbf{R}_2\, \mathbf{T}(\mathbf{R}_1 - \mathbf{R}_2)\mathbf{T}(\mathbf{R}_2 - \mathbf{R}_1) \\ - \zeta_1^2\zeta_2^2 n_2^2 \int d\mathbf{R}_2 d\mathbf{R}_3\, \mathbf{T}(\mathbf{R}_1 - \mathbf{R}_2)\mathbf{T}(\mathbf{R}_2 - \mathbf{R}_3)\mathbf{T}(\mathbf{R}_3 - \mathbf{R}_1) + \ldots\end{aligned} \tag{7.8}$$

where $n_2 = N_2/\Omega$, and N_2 is the number of the fixed particles in the system and Ω is a system volume.

In the Fourier space transform we have

$$\mathbf{T}(\mathbf{k}) = (\mathbf{U} - \mathbf{kk}/k^2)/\eta k^2$$

and the series (7.8) can be represented using the convolution theorem in the form

$$\begin{aligned}\overline{\xi_{11}} &= \zeta_1 + \zeta_1^2\zeta_2 n_2 \int \frac{d\mathbf{k}}{(2\pi)^3}\left(\mathbf{U} - \frac{\mathbf{kk}}{k^2}\right) \\ &\quad \times \left[\left(\frac{1}{\eta k^2}\right)^2 - \zeta_2 n_2 \left(\frac{1}{\eta k^2}\right)^3 + \zeta_2^2 n_2^2 \left(\frac{1}{\eta k^2}\right)^4 - \ldots\right] \\ &= \zeta_1 + \zeta_1^2\zeta_2 n_2 \int \frac{d\mathbf{k}}{(2\pi)^3} \frac{1/\eta k^2}{\eta k^2 + \zeta_2 n_2}\end{aligned} \tag{7.9}$$

Here we have used the well–known formula for the sum of the geometric progression appearing in square brackets. Calculating the integral, we find that the drag coefficient is the diagonal tensor with coefficients

$$\overline{\xi_{11}} = \zeta_1\left(1 + 3\frac{a_1}{a_2}\sqrt{\frac{\varphi_c}{2}}\right) \tag{7.10}$$

Recall that this result has been obtained in the loop graph approximation. Now we expand the class of summing graphs, including graphs with coinciding indexes. Then, as an example, the additional graph

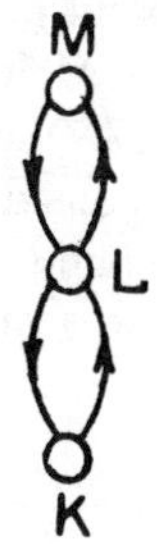

(7.11)

appears in the fourth order term. This graph describes the propagation of the perturbation generated by the Kth particle through the Lth particle to the Mth particle and, reverse, running through the same Lth particle to the Kth particle.

Three graphs are added to the loop graph of the fifth order; they are

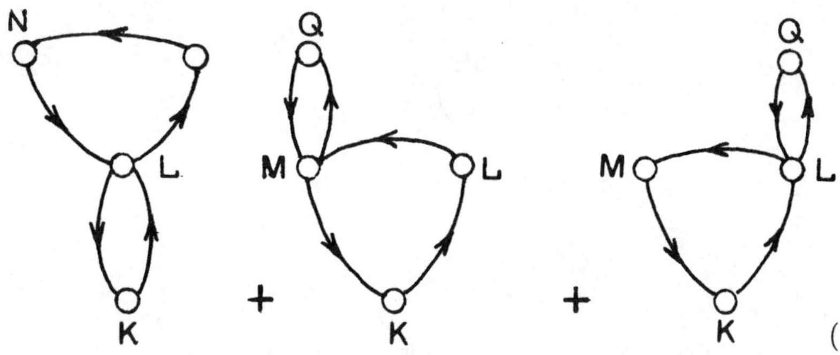

(7.12)

Taking into account all the graphs of this type, in all terms of the power expansion (7.6) results in the series

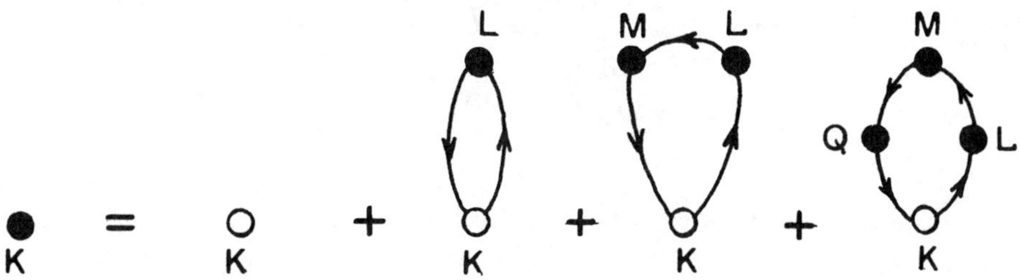

(7.13)

As a real example, the graph (7.11) and the first one of equation (7.12) together the similar graphs from the terms of higher orders provide a renormalization of the Lth vortex of the second graph in equation (7.13), that is, the change of ζ_L on $\overline{\xi_{LL}}$. The second and third graphs in equation (7.12) contribute to renormalization of the Mth and Lth vertices of the third term in the right–hand side of equation (7.7), etc. Then equation (7.13) becomes the equation in $\overline{\xi_{KK}}$. At first we consider it for a fixed particle ($K \neq 1$). Neglecting any influence of the suspended particle (which gives at most an error of the order of $\zeta_1/N2\zeta_2$), we rewrite equation (7.13) in the algebraic form

$$\begin{aligned}\overline{\xi_{22}} &= \zeta_2 + \overline{\xi_{22}}\zeta_2^2 n_2 \int d\mathbf{R}_3\, \mathbf{T}(\mathbf{R}_2 - \mathbf{R}_3)\mathbf{T}(\mathbf{R}_3 - \mathbf{R}_2) \\ &\quad - \overline{\xi_{22}}\zeta_2^2 n_2^2 \int d\mathbf{R}_3 d\mathbf{R}_4 \mathbf{T}(\mathbf{R}_2 - \mathbf{R}_3)\mathbf{T}(\mathbf{R}_3 - \mathbf{R}_4)\mathbf{T}(\mathbf{R}_4 - \mathbf{R}_2) + \ldots \end{aligned} \tag{7.14}$$

The analysis of this equation is precisely the same as that of equation (7.8) and as a result, we get

$$\overline{\xi_{22}} = \zeta_2 + \overline{\xi_{22}}\zeta_2^2 n_2 \int \frac{d\mathbf{k}}{(2\pi)^3} \left(\mathbf{U} - \frac{\mathbf{kk}}{k^2}\right) \frac{1}{\eta k^2 + \overline{\xi_{22}} n_2} = \zeta_2 \left(1 + a_2 \sqrt{\frac{n_2 \overline{\xi_{22}}}{\eta}}\right) \tag{7.15}$$

The solution to this equation is

$$\overline{\xi_{22}} = \zeta_2 \left(1 + 3\sqrt{\frac{\varphi_2}{2}} + \frac{9}{2}\varphi_2\right) \simeq \zeta_2 \left(1 + 3\sqrt{\frac{\varphi_2}{2}}\right) \tag{7.16}$$

Now we return to equation (7.13) for the case $K = 1$. Then it follows, from equation (7.8) by changing ζ_2 to $\overline{\xi_{22}}$ in all terms. Then in Fourier transform we get

$$\begin{aligned}\overline{\xi_{11}} &= \zeta_1 + \zeta_1^1 \overline{\xi_{22}} n_2 \int \frac{d\mathbf{k}}{(2\pi)^3} \left(U - \frac{\mathbf{kk}}{k^2}\right) \frac{1/\eta k^2}{\eta k^2 + \overline{\xi_{22}} n_2} \\ &= \zeta_1 \left(1 + a_1 \sqrt{\frac{n_2 \overline{\xi_{22}}}{\eta}}\right)\end{aligned} \tag{7.17}$$

which coincides with equation (7.10) in the lowest approximation in φ_2.

Thus, we have ensured that in the approximation under consideration, taking into account the renormalizing vertices in the graph expansion (7.13) does not affect the value of the coefficient $\overline{\xi_{11}}$.

Averaging equation (7.4) over the coordinates of all the fixed particles, it is easy to verify that in the lowest order of φ_2, the mean perturbation of the flow velocity is actually determined by the expression in equation (5.8). Therefore the correlation function $\Gamma_{xx}(\mathbf{k})$ in (6.11) can be found in terms of the series (7.4). The averaged product of this series is

$$\begin{aligned}\overline{\mathbf{v}(\mathbf{r})\mathbf{v}(\mathbf{r}')} = \{[&-\zeta_2 \sum_i \mathbf{T}_{ri} + \zeta_2^2 \sum_{ij} \mathbf{T}_{rj}\mathbf{T}_{ji} + \zeta_2^3 \sum_{ijk} \mathbf{T}_{rk}\mathbf{T}_{kj}\mathbf{T}_{ji} + \ldots] \\ &\times[-\zeta_2 \sum_l \mathbf{T}_{r'l} + \zeta_2^2 \sum_{lm} \mathbf{T}_{r'm}\mathbf{T}_{ml} \\ &\quad -\zeta_2^3 \sum_{lmn} \mathbf{T}_{r'n}\mathbf{T}_{nm}\mathbf{T}_{ml} + \ldots]\}_{av} : \mathbf{v}_0\mathbf{v}_0\end{aligned} \tag{7.18}$$

where the braces in the expression $\{\Phi\}_{av}$ are equivalent to the bar over Φ. By multiplying the terms in square brackets, we can represent (7.18) in the form

$$\begin{aligned}\overline{\mathbf{v}(\mathbf{r})\mathbf{v}(\mathbf{r}')} &= (\lambda_1 + \lambda_2 + \lambda_3 + \ldots) : \mathbf{v}_0\mathbf{v}_0 \\ \lambda_1 &= \frac{\zeta_2^2}{\Omega} \sum_i \int d\mathbf{R}_i\, \mathbf{T}_{ri} \left[\mathbf{T}_{r'i} - \frac{\zeta_2}{\Omega} \sum_m \int d\mathbf{R}_m\, \mathbf{T}_{r'm}\mathbf{T}_{mi} + \ldots\right] \\ \lambda_2 &= \frac{\zeta_2^2}{\Omega^2} \sum_{il(i\neq l)} \int d\mathbf{R}_i d\mathbf{R}_l\, \mathbf{T}_{rl} \left[T_{r'l} - \frac{\zeta_2}{\Omega} \sum_m \int d\mathbf{R}_m\, \mathbf{T}_{r'm}\mathbf{T}_{ml} + \ldots\right] \\ \lambda_3 &= \frac{\zeta_2^3}{\Omega^2} \sum_{ij} \int d\mathbf{R}_i d\mathbf{R}_j\, \mathbf{T}_{rj}\mathbf{T}_{ji} \Big[\mathbf{T}_{r'i} \\ &\quad -\zeta_2 \mathbf{T}_{r'j}\mathbf{T}_{ji} - \frac{\zeta_2}{\Omega} \sum_{m(\neq j)} \int d\mathbf{R}_m \mathbf{T}_{r'm}\mathbf{T}_{mi} + \ldots\Big]\end{aligned} \tag{7.19}$$

The expression for λ_1 corresponds to the product of the first term from the first square brackets in (7.18) by all terms from the second square brackets with $l = i$, and is of order n and the expression for λ_2 with $l \neq i$ is of order n^2. Therefore, for small n, we can ignore the contribution of λ_2. Physically, it means that the basic contribution to the correlation function $\Gamma(\mathbf{r} - \mathbf{r}')$ is that of the disturbance generated in the same ith particle (with subsequent summation over all i).

The second term in λ_3 contains the factor $\mathbf{T}_{ji}$, identical to that preceding the square bracket. Both describe the propagation of the perturbation from the ith to the jth particle. Therefore the term with $\mathbf{T}_{ij}$ can simply be dropped, since it has a higher order of smallness with respect to a_2/R_{ij}, compared with the first term in λ_3. For the same reason, in all subsequent $\lambda_n (n > 3)$ we can omit terms that are the products from different square brackets with identical indexes in (7.18). Then (7.19) takes the form

$$\begin{aligned}\overline{\mathbf{v}(\mathbf{r})\mathbf{v}(\mathbf{r}')} &= \zeta_2^2 n_2 \int d\mathbf{R}_2 \left[\mathbf{T}_{r2} - \zeta_2 n_2 \int d\mathbf{R}_3\, \mathbf{T}_{r3}\mathbf{T}_{32} \right.\\ &\qquad \left. + \zeta_2^2 n_2^2 \int d\mathbf{R}_3 d\mathbf{R}_4\, \mathbf{T}_{r4}\mathbf{T}_{43}\mathbf{T}_{32} - \dots\right] \\ &\quad \times \left[\mathbf{T}_{r'2} - \zeta_2 n_2 \int d\mathbf{R}_3\, \mathbf{T}_{r'3}\mathbf{T}_{32} \right.\\ &\qquad \left. + \zeta_2^2 n_2^2 \int d\mathbf{R}_3 d\mathbf{R}_4\, \mathbf{T}_{r'4}\mathbf{T}_{43}\mathbf{T}_{32} - \dots\right] : \mathbf{v}_0\mathbf{v}_0\end{aligned} \tag{7.20}$$

or, in the Fourier representation,

$$\begin{aligned}\Gamma_{xx}(\mathbf{k}) &= \zeta_2^2 n_2 v_0^2 \left[\mathbf{T}(\mathbf{k}) - \zeta_2 n_2 (T(\mathbf{k}))^2 + \zeta_2^2 n_2^2 (\mathbf{T}(\mathbf{k}))^3 - \dots\right]_{xz}^2 \\ &= \zeta_2^2 n_2 v_0^2 \left(\frac{1}{\eta k^2}\right)^2 \frac{k_x^2 k_z^2}{k^4(1+\zeta_2 n_2/(\eta k^2))^2} \\ &= (6\pi a_2 v_0)^2 n_2 \frac{k_x^2 k_z^2}{k^4(k^2+\Lambda^2)^2} \\ &\left(\Lambda = \sqrt{\zeta_2 n_2/\eta} = \sqrt{6\pi a_2 n_2}\right)\end{aligned} \tag{7.21}$$

where Λ is the inverse length of the screening in a dispersion system (in the approximation of point particles).

On substituting expressions (6.12) and (7.21) into (6.13), we obtain

$$\Delta D_{\perp} = \frac{(6\pi a_2 v_0)^2 n_2}{(2\pi)^3} \int d\mathbf{k} \frac{k_x^2 k_z^2}{k^4(k^2+\Lambda^2)^2(Dk^2 + iv_0 k_z)} \tag{7.22}$$

On evaluating this integral, we find that

$$\begin{aligned}\Delta D_{\perp} &= \frac{9}{2}\pi a_2^2 D n_2 \Lambda^{-1} \left[\frac{1}{3} + \alpha(1-2\alpha) - \alpha(1-2\alpha^2)\ln\frac{\alpha+1}{\alpha}\right] \\ \alpha &= \frac{\Lambda D}{v_0}\end{aligned} \tag{7.23}$$

In the limiting case $\alpha \ll 1$, this expression takes the form

$$\Delta D_{\perp} = \frac{3}{2}\pi a_2^2 D n_2 \Lambda^{-1} = \frac{3}{4}\sqrt{\frac{\varphi_2}{2}} D \tag{7.24}$$

It follows from (6.13) and (7.24) that

$$D_\perp = D\left(1 + \frac{3}{4}\sqrt{\frac{\varphi_2}{2}}\right) \qquad (7.25)$$

The coefficient D also depends on the concentration of fixed particles. In fact, by definition of D (5.9) and expression (7.10), we have

$$D = D_0\left(1 - 3\frac{a_1}{a_2}\sqrt{\frac{\varphi}{2}}\right), \quad D_0 = \frac{kT}{6\pi a_1 \eta} \qquad (7.26)$$

where D_0 is the diffusion coefficient of a particle of radius a_1 in a pure fluid with viscosity η. Thus, finally, from (7.25) and (7.26) we obtain

$$D_\perp = D_0\left[1 - 3\left(\frac{a_1}{a_2} - \frac{1}{4}\right)\sqrt{\frac{\varphi_2}{2}}\right] \qquad (7.27)$$

Hence it is seen that, depending on the ratio a_1/a_2, the diffusion coefficient $D_\perp$ can be larger or smaller than the Brownian diffusion coefficient D_0 in a pure fluid. This is due to the change in the relative contribution of two competing influences of fixed particles on the diffusion of a Brownian particle. On one hand, their presence increases the resistance to the motion of particle $\overline{\xi_{11}}$, and correspondingly reduces the diffusion coefficient (7.26). On the other hand, in a flow past a randomly configurated grid, there a random velocity fluid appears, which leads to an additional fluctuation motion (relative to the thermal motion) of the particle and, correspondingly, to an increase in the diffusion coefficient.

The calculation of the coefficient $D_\perp$ in the system described was also discussed in [83]. However, the result obtained there is not exactly accurate since the authors did not take into account the difference between D and D_0, given by (7.26). Also, in computing $\Delta D_\perp$, they used an incorrect expression for $\Gamma_{xx}(\mathbf{k})$ in the form of a product $n_2 u_x(\mathbf{k}) v_x(-\mathbf{k})$ where $\mathbf{u}(\mathbf{k})$ is the Fourier transform of a non-screened velocity field of one mobile particle. Such a choice of $\Gamma_{xx}(\mathbf{k})$ corresponds to allowing for only the first term of λ_1 in expansion (7.19), or the first term in the first square brackets in (7.20). In computing $\Delta D_\perp$, this leads to a double result compared with (7.23). Further, the authors made a mistake in evaluating the integral for $\Delta D_\perp$, and their correction of the transverse–diffusion coefficient as calculated was four times as great as that given by equation (7.23) in the present paper.

The result obtained can be used in the theory of the marker expanding in a fixed bed. This problem is of great importance in chemical engineering where one of the most spread methods of the studies of the catalyst fixed bed is based on detections of behavior of a colored, radioactive, or thermal marker upon its motion through a fixed bed in rising fluid flow.

It should be noted that if we introduce the dimensionless parameter the Peclet number, $\mathrm{Pe} = v_0 a/D_0$, characterizing a relative intensity of diffusion and convective transport, then the results obtained are referred to the limiting case of low Peclet number, $\mathrm{Pe} \ll 1$. There has been no sufficiently rigorous statistical mechanical approach to the problem at intermediate ($\mathrm{Pe} \sim 1$) or large ($\mathrm{Pe} \gg 1$) values of the Peclet number, so one must be satisfied with phenomenological approaches in this range of Peclet number values (see, for example, [84, 85]) .

Appendix A

Vector and Tensor Divergencies in Curvilinear Coordinates

The system of coordinates related to a surface Σ consists of the axis z directed along the normal $\mathbf{n}$ to the surface Σ, and curvilinear coordinates q_1 and q_2 pertaining to Σ.

The divergence of a three-dimensional vector $\mathbf{a}$ can be represented in these coordinates in the form

$$\begin{aligned}\operatorname{div}\mathbf{a} = {} & \frac{\partial a_z}{\partial z} + a_z\left(\frac{1}{H_1}\frac{\partial H_1}{\partial z} + \frac{1}{H_2}\frac{\partial H_2}{\partial z}\right) + \frac{1}{H_1}\frac{\partial a_1}{\partial q_1} \\ & + \frac{1}{H_2}\frac{\partial a_2}{\partial q_2} + \frac{1}{H_1 H_2}\left(a_1\frac{\partial H_2}{\partial q_1} + a_2\frac{\partial H_1}{\partial q_2}\right)\end{aligned} \tag{A.1}$$

where H_1 and H_2 are the Lame parameters. It is noted here, that

$$\frac{1}{H_1}\frac{\partial H_1}{\partial z} = \frac{1}{R_1}, \qquad \frac{1}{H_2}\frac{\partial H_2}{\partial z} = \frac{1}{R_2}$$

where R_1 and R_2 are the radii of curvature.

Equation (A.1) can be rewritten as

$$\operatorname{div}\mathbf{a} = \frac{\partial a_z}{\partial z} + \operatorname{div}_\tau \mathbf{a} \tag{A.2}$$

or

$$\operatorname{div}\mathbf{a} = \frac{\partial a_z}{\partial z} + a_z\left(\frac{1}{R_1} + \frac{1}{R_2}\right) + \operatorname{div}_\tau \mathbf{a}_\tau \tag{A.3}$$

where $\mathbf{a}_\tau$ is the two–dimensional vector tangential to the surface Σ.

Equations (A.1)–(A.3) define the two–dimensional divergence operator acting on three– and two–dimensional vectors.

In the particular case of a spherical surface Σ, we have $z = R$, $q_1 = \vartheta$, $q_2 = \varphi$, $H_1 = R$, $H_2 = \sin\vartheta$,

$$\operatorname{div}\mathbf{a} = \frac{\partial a_R}{\partial R} + \frac{2}{R}a_R + \frac{1}{R}\frac{\partial a_\vartheta}{\partial \vartheta} + \frac{1}{R\sin\vartheta}\frac{\partial a_\varphi}{\partial \varphi} + \frac{a_\vartheta}{R\sin\vartheta}\frac{\partial \sin\vartheta}{\partial \vartheta} \tag{A.4}$$

$$\operatorname{div}\mathbf{a}_\tau = \frac{2}{R}a_R + \operatorname{div}_\tau \mathbf{a}_\tau \tag{A.5}$$

$$\operatorname{div}_\tau \mathbf{a}_\tau = \frac{1}{R}\frac{\partial a_\vartheta}{\partial \vartheta} + \frac{1}{R\sin\vartheta}\frac{\partial a_\varphi}{\partial \varphi} + \frac{a_\vartheta}{R\sin\vartheta}\frac{\partial \sin\vartheta}{\partial \vartheta} \tag{A.6}$$

A divergence of a three–dimensional tensor $\mathbf{T}$ is a vector. Its components in curvilinear coordinates z, q_1, and q_2 are

$$\begin{aligned}(\operatorname{Div}\mathbf{T})_z &= \frac{\partial T_{zz}}{\partial z} + \Big(T_{zz} - T_{11}\Big)\frac{1}{H_1}\frac{\partial H_1}{\partial z} + \Big(T_{zz} - T_{22}\Big)\frac{1}{H_2}\frac{\partial H_2}{\partial z} \\ &+ \frac{1}{H_1 H_2}\Big[\frac{\partial}{\partial q_1}\Big(H_2 T_{1z}\Big) + \frac{\partial}{\partial q_2}\Big(H_1 T_{2z}\Big)\Big] = \frac{\partial T_{zz}}{\partial z} + (\operatorname{Div}_\tau T)_z \\ &= \frac{\partial T_{zz}}{\partial z} + \Big(T_{zz} - T_{11}\Big)\frac{1}{R_1} + \Big(T_{zz} - T_{22}\Big)\frac{1}{R_2} + Div_\tau T_{\tau z}\end{aligned} \tag{A.7}$$

$$\begin{aligned}(\operatorname{Div}\mathbf{T})_1 &= \frac{1}{H_1 H_2}\Big[\frac{\partial}{\partial q_1}\Big(H_2 T_{11}\Big) + \frac{\partial}{\partial q_2}\Big(H_1 T_{21}\Big) + \frac{\partial}{\partial z}\Big(H_1 H_2 T_{z1}\Big)\Big] \\ &+ \Big[\frac{T_{12}}{H_1 H_2}\frac{\partial H_1}{\partial q_2} + \frac{T_{1z}}{R_1} - \frac{T_{22}}{H_1 H_2}\frac{\partial H_2}{\partial q_1}\Big]\end{aligned} \tag{A.8}$$

$$\begin{aligned}(\operatorname{Div}\mathbf{T})_2 &= \frac{1}{H_1 H_2}\Big[\frac{\partial}{\partial q_1}\Big(H_2 T_{12}\Big) + \frac{\partial}{\partial q_2}\Big(H_1 T_{22}\Big) + \frac{\partial}{\partial z}\Big(H_1 H_2 T_{z2}\Big)\Big] \\ &+ \Big[\frac{T_{21}}{H_1 H_2}\frac{\partial H_2}{\partial q_1} + \frac{T_{2z}}{R_2} - \frac{T_{11}}{H_1 H_2}\frac{\partial H_1}{\partial q_2}\Big]\end{aligned} \tag{A.9}$$

The two–dimensional divergence of a three–dimensional tensor is introduced in equation (A.7). In the particular case of a diagonal tensor, that is, when $T_{11} = T_{22} = T_{\tau\tau}, T_{zz} \neq T_{\tau\tau}$, this equation takes the form

$$(\operatorname{Div}\mathbf{T})_z = \frac{\partial T_{zz}}{\partial z} + \Big(T_{zz} - T_{\tau\tau}\Big)\Big(\frac{1}{R_1} + \frac{1}{R_2}\Big) \tag{A.10}$$

Appendix B

Mechanical Definition of Surface Tension

There is a common procedure [15] for defining the surface tension coefficient via the free energy of a two–phase system containing the dividing surface. Such a definition is the most general in a thermodynamically equilibrium state, but it is not valid for nonequilibrium situations.

In this case the surface tension should be introduced without use of thermodynamic functions, in favor of pure mechanical conceptions of work and pressure.

For simplicity the discussion will be confined to a steady state case. We consider an unbounded rectangular tube containing bulk phases **0** and **I** (see Figure B.1). They are divided by a syrface layer filling the volume between the surfaces Σ_0 and Σ_1.

Pressure anisotropy is inherent in a surface layer, so it is possible to introduce the normal pressure $p_n(z)$ along the normal to the layer, and tangential pressure $p_\tau(z)$.

The volume V of the surface layer can be expanded on $dV = dzdA$ by way of withdrawal of a narrow strip of the tube wall of height dz; dA is the bottom area of the withdrawn volume. Expanding the system produces the work $dW_1(z) = p_\tau(z)dV$. As thickness and density of the surface layer are determined by physical properties of the system and do not depend on the geometry of the tube; the expansion of the surface layer volume on dV is accompanied by the fluid inflow into the surface layer from the bulk phases. While this inflow is going on, the work $dW_2(z) = -p_n(z)dV$ is produced. The summary work is

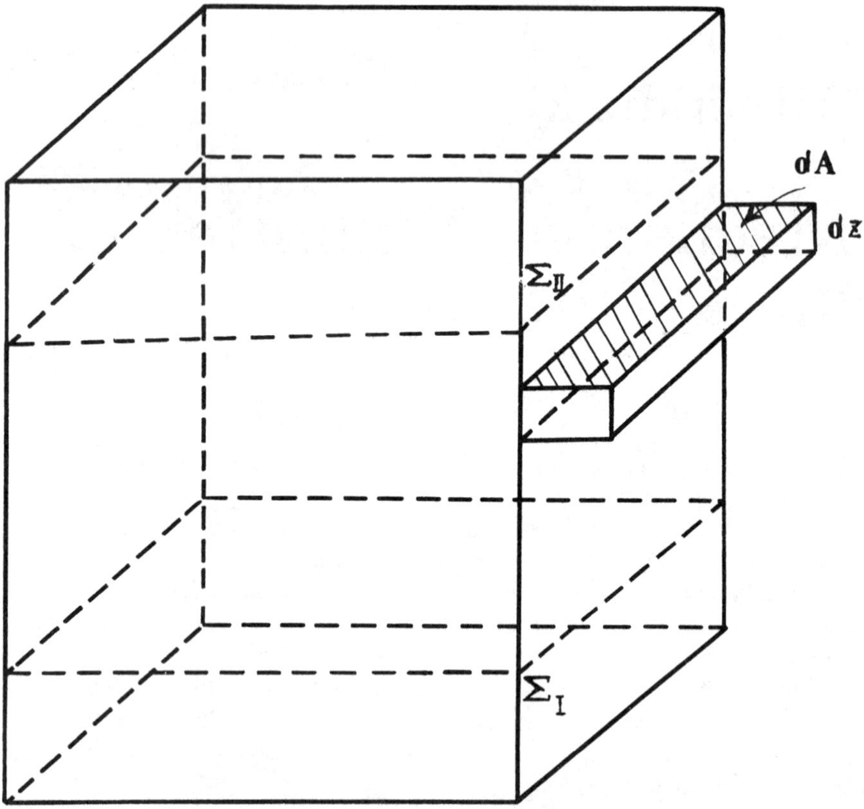

Figure B.1: The rectangular tube containing bulk phases **0** and **I** divided by a syrface layer filling the volume between the surfaces Σ_0 and Σ_1. The surface layer is expanded on $dV = dzdA$. (To derivation of equation (B.1).) coefficient

$$dW(z) = dW_1(z) + dW_2(z) = (p_\tau(z) - p_n(z))dV$$

When the whole of the surface layer is expanded on the area dA, the work is

$$\Delta W = \int_{\Sigma_0}^{\Sigma_1} dz(p_\tau(z) - p_n(z)) \cdot dA = -\gamma dA$$

where

$$\gamma = \int_{-\infty}^{+\infty} dz(p_\tau(z) - p_n(z)) \tag{B.1}$$

is the surface tension coefficient. Here the integration limits are expanded up to $\pm\infty$; this does not influence the value γ, as the pressure is isotropic outside the surface layer, and the integrand is zero there. The sign of the definition of γ is chosen such that it would be positive for an equilibrium stabile interphase boundary.

In an equilibrium system the pressure anisotropy is due to an anisotropy of a mean virial of an intermolecular interaction potential. This is why it is widely accepted (see, as an example, [86]) that surface tension is related entirely to uncompensated intermolecular interactions in a surface layer.

Conversely, in a nonequilibrium system, a pressure can be anisotropic even in the absence of intermolecular interactions. If the temperature is anisotropic in a layer of discontinuity of an ideal gas flow, then the normal and tangential pressures are

$$p_n(z) = \varrho(z)RT_n(z), \qquad p_\tau(z) = \varrho(z)RT_\tau(z)$$

where $\varrho(z)$ is the gas density profile, and R is the gas constant.

On the other hand, the pressure anisotropy is interpreted in fluid dynamics as a viscous stress tensor, so one may say that the surface tension of a gas discontinuity is related to viscous stresses within the discontinuity layer.

Appendix C

Exclusion of Time–Derivatives of Parameters

Here the practice of excluding time–derivatives for bulk phases will be discussed, aqs in the monograph [4]. Hereafter the upper indexes of bulk phase parameters **0**, **I** are omitted for brevity.

For $F_1(\mathbf{x},t) = \beta(\mathbf{x},t)$ we have

$$\frac{\partial \beta}{\partial t} = \frac{\partial \beta}{\partial u}\frac{\partial u}{\partial t} + \sum_{\alpha} \frac{\partial \beta}{\partial \varrho_\alpha}\frac{\partial \varrho_\alpha}{\partial t} \tag{C.1}$$

Introducing the notation $\frac{d}{dt} = \frac{\partial}{\partial t} + \mathbf{v} \cdot \nabla$ and taking into account equations (2.5.10) and (2.5.13), this equation can be rewritten as

$$\begin{aligned} \frac{d\beta}{dt} &= \left\{ (u+p)\frac{\partial \beta}{\partial u} - \sum_{\alpha} \frac{\partial \beta}{\partial \varrho_\alpha}\varrho_\alpha \right\} \operatorname{div} \mathbf{v} \\ &= \left\{ \frac{\partial p}{\partial \beta}\frac{\partial \beta}{\partial u} - \sum_{\alpha} \frac{\partial p}{\partial \nu_\alpha}\frac{\partial \beta}{\partial \varrho_\alpha} m_\alpha \right\} \beta \operatorname{div} \mathbf{v} \end{aligned} \tag{C.2}$$

Here we used the thermodynamic equations

$$\frac{\partial p}{\partial \beta} = -\frac{u+p}{\beta}, \frac{\partial p}{\partial \nu_\alpha} = \frac{\varrho_\alpha}{m_\alpha \beta} \tag{C.3}$$

that follow from the Gibbs–Duhem equation (2.3.9).

Notice that in accordance with equation (2.3.8) we have

$$\beta = \frac{\partial S}{\partial u}, \qquad -\nu_\alpha = \frac{\partial S}{\partial n_\alpha} \tag{C.4}$$

where S is the entropy density as a function of u and $n_\alpha = \varrho_\alpha/m_\alpha$. Therefore, the parameters β and ν_α are related in the following way:

$$m_\alpha \frac{\partial \beta}{\partial \varrho_\alpha} = -\frac{\partial \nu_\alpha}{\partial u} = \frac{\partial^2 S}{\partial u \partial n_\alpha} \tag{C.5}$$

So the equation (C.2) can be rewritten in the form

$$\begin{aligned} \frac{d\beta}{dt} &= \Big\{ \frac{\partial p}{\partial \beta}\frac{\partial \beta}{\partial u} + \sum_\alpha \frac{\partial p}{\partial \nu_\alpha}\frac{\partial \nu_\alpha}{\partial u} \Big\} \beta \operatorname{div} \mathbf{v} \\ &= \beta \Big(\frac{\partial p}{\partial u}\Big)_n \operatorname{div} \mathbf{v} \end{aligned} \tag{C.6}$$

or

$$\frac{\partial \beta}{\partial t} = -\mathbf{v}\nabla\beta + \beta\Big(\frac{\partial p}{\partial u}\Big)_n \operatorname{div} \mathbf{v} \tag{C.7}$$

Thus, the time derivatives of $\beta^{0,1}$ are determined in terms of $\nabla\beta^{0,1}$ and $\operatorname{div} \mathbf{v}^{0,1}$.

The time derivative of the parameter $F_2 = -\beta\mathbf{v}$ incorporates the derivative $\frac{\partial \beta}{\partial t}$, as well as the time derivative of the velocity $\mathbf{v}$, which is determined from the Euler equation (2.5.12). It is sufficient to assume that pressure, p, is a local function of β and ν_α. Then taking into account (C.3), we get

$$\varrho \frac{\partial \mathbf{v}}{\partial t} = -\varrho \mathbf{v} \cdot \nabla \mathbf{v} + \frac{u+p}{\beta}\nabla\beta - \sum_\alpha \frac{\varrho_\alpha}{m_\alpha \beta}\nabla \nu_\alpha \tag{C.8}$$

Thus the time derivative of velocity is expressed in terms of the gradients $\nabla\beta$, $\nabla\nu_\alpha$, and $\nabla\mathbf{v}$.

The time derivative of the parameter $F_{3\alpha} = -\nu_\alpha + \beta m_\alpha v^2/2$ incorporates the above–determined derivatives of β and $\mathbf{v}$, and also, the time derivative $\partial \nu_\alpha/\partial t$. Supposing $\nu_\alpha(\mathbf{x})$ is a local function of $u(\mathbf{x})$ and $\varrho_1(\mathbf{x}), ..., \varrho_\beta(\mathbf{x})$, using equations (2.5.10) and (2.5.13), we get

$$\begin{aligned} \frac{d\nu_\alpha}{dt} &= \frac{\partial \nu_\alpha}{\partial u}\frac{du}{dt} + \sum_\beta \frac{\partial \nu_\alpha}{\partial \varrho_\beta}\frac{d\varrho_\beta}{dt} \\ &= -\frac{\partial \nu_\alpha}{\partial u}(u+p)\operatorname{div}\mathbf{v} - \sum_\beta \frac{\partial \nu_\alpha}{\partial \varrho_\beta}\varrho_\beta \operatorname{div}\mathbf{v} \end{aligned} \tag{C.9}$$

Taking into account (C.3), and thermodynamic relations

$$\frac{\partial \nu_\alpha}{\partial \varrho_\beta} = \frac{\partial \nu_\beta}{\partial \varrho_\alpha} \tag{C.10}$$

which can be derived by the same way as equation (C.5), we get

$$\frac{d\nu_\alpha}{dt} = \left\{ \frac{\partial p}{\partial \beta} \frac{\partial \beta}{\partial \varrho_\alpha} m_\alpha + \sum_\beta \frac{\partial p}{\partial \nu_\beta} \frac{\partial \nu_\beta}{\partial \varrho_\alpha} \right\} \beta \operatorname{div} \mathbf{v} \tag{C.11}$$

or

$$\frac{\partial \nu_\alpha}{\partial t} = -\mathbf{v}\nabla \nu_\alpha - \beta m_\alpha \left(\frac{\partial p}{\partial \varrho_\alpha} \right)_u \operatorname{div} \mathbf{v} \tag{C.12}$$

Thus the time derivative of ν_α is determined in the terms of $\nabla \nu_\alpha$ and $\operatorname{div} \mathbf{v}$.

Bibliography

[1] 1. **N. N. Bogoliubov**. *Problems of Dynamical Theory in Statistical Physics.* ONTI, Moscow, 1946. (in Russian) [English translation in Studies in Statistical Mechanics, vol.1, North–Holland Publ.Co., Amsterdam, 1962].

[2] 2. **H. Mori**. "Statistical–Mechanical Theory of Transport Processes in Fluids". *Phys. Rev.*, 112:1829–1843, 1958.

[3] 3. **R. Balescu**. *Equilibrium and Nonequilibrium Statistical Mechanics.* John Wiley, New York, 1975.

[4] 4. **D. N. Zubarev**. *Nonequilibrium Statistical Thermodynamics.* Plenum Press, New York, 1974.

[5] 5. **N. N. Bogoliubov**. Quasiaverages in problems of statistical mechanics. Preprint D–781, UINI, Dubna, 1961. (in Russian).

[6] 6. **R. Luzzi and A. R. Vasconcellos**. "On the Nonequilibrium Statistical Operator Method". *Forschr. Phys.*, 38(11):887–922, 1990.

[7] 7. **S. R. de Groot and P. Mazur**. *Nonequilibrium Thermodynamics.* North–Holland Publ. Co., Amsterdam, 1962.

[8] 8. **L. Waldmann**. "Non-equilibrium thermodynamics of boundary conditions". *Z. Naturforsch.*, Bd.22a:1269–1281, 1967.

[9] 9. **D. Bedeaux and A. M. Albano and P. Mazur**. "Boundary conditions and non-equilibrium thermodynamics". *Physica*, 82A:438–451, 1976.

[10] 10. **D. Bedeaux**. "Nonequilibrium Thermodynamics and Statistical Physics of Surfaces". *Adv. Chem. Phys.*, 64:47–109, 1986.

[11] 11. **J. Kovac**. "Non-equilibrium thermodynamics of interfacial systems". *Physica*, 86A:1–22, 1977.

[12] 12. **F. C. Goodrich**. "The Theory of Capillary Excess Viscosities". *Proc. Roy. Soc., London*, A374:341–370, 1981.

[13] 13. **J. S. Rowlinson, and B. Widom**. *Molecular Theory of Capillarity.* Clarendon Press, Oxford, 1982.

[14] 14. **A. Messiah**. *Quantum Mechanics*, Volume 1. North–Holland Publ. Co., Amsterdam, 1961.

[15] 15. **S. Ono and S. Kondo**. "Molecular Theory of Surface Tension in Liquids". *Handbuch der Physik*, 10:134–208, 1960.

[16] 16. **F. C. Goodrich**. "Surface Viscosity as a Capillary Excess Transport Property". In A. I. Rusanov and F. C. Goodrich, eds, *The Modern Theory of Capillarity*, p. 38–61. Khimia Publisher, Leningrad, 1980.

[17] 17. **J. Boussinesq**. "Sur l'existence d'une viscjsite superficielle". *Ann. Chim. Phys.*, 39:349–357, 1913.

[18] 18. **L. E. Scriven**. "Dynamics of a Fluid Interface". *Chem. Engn. Sci.*, 12(1):98–108, 1960.

[19] 19. **F. C. Goodrich**. "The Mathematical Theory of Capillarity". *Proc. Roy. Soc., London*, A260:490–509, 1960.

[20] 20. **V. V. Shuleykin**. *Physics of Sea.* Nauka, 1968. (in Russian).

[21] 21. **A. Bashkirov and G. Korol'kov**. "On the Surface Viscosity at the Boundary between Phases". *PMM Journ. Applied Math. and Mech. (USSR)*, 47(6):832–835, 1983.

[22] 22. **V. G. Levich**. *Physicochemical Hydrodynamics.* Prentice–Hall, Englewood Cliffs, NJ, 1962.

[23] 23. **J. C. Earnshaw**. "Surface Viscosity of Water". *Nature*, 292:138–139, 1981.

[24] 24. **A. G. Bashkirov**. "Influence of Viscous Stresses on Shock Wave Stability in Gases". . *Phys. Fluids A*, 3(5):960–966, 1991.

[25] 25. **D. L.Chubb**. "Ionizing Shock Structure in a Monatomic Gases". *Phys. Fluids*, 11(11):2363–2376, 1968.

[26] 26. **T. G. Pai and S. M. Ramachandra**. "Structure of a Shock Wave in a Vibrationally Relaxing Diatomic Gas". *Phys. Fluids*, 18(2):166–174, 1975.

[27] 27. **A. G. Bashkirov and A. V. Orlov**. "Transport Phenomena in a Plane Shock Wave". *Journ. Statistical Physics*, 64:429–436, 1991.

[28] 28. **G. A. Bird**. "Aspects of the Structure of a Strong Shock Waves". *Phys. Fluids*, 13(5):1172–1177, 1970.

[29] 29. **E. Muntz and L. M. Harnett**. "Molecular Velocity Distribution Functions Measurment in a Normal Shock Wave". *Phys. Fluids*, 12(10):2027–2035, 1969.

[30] 30. **W. M. Kornegay and J. D. Fridman and W. C. Worthington**. "Studies of Spherical Shock Waves at Law Ambient Pressure". In *Proceedings of 6-th Intern. Symp. on Rarefied Gas Dynamics*, Volume 1, p. 863–870, 1969.

[31] 31. **S. P. Dyakov**. "On Shock Wave Stability". *Zh. Exp. Teor. Fyz.*, 27(3):288–295, 1954 (in Russian), [Air Force Office of Science Res. No.AFOSR- TN-56-406, 1956 (English Translation)].

[32] 32. **S. V. Iordanski**. "On the Stability of Plane Stationary Shock Wave". *Prikl. Matem. Mehanika*, 21(4):463–472, 1957 (in Russian).

[33] 33. **V. M. Kontorovitz**. "To the Question of the Shock Wave Stability". *Zh. Exp. Teor. Fyz.*, 33(9):1525–1527, 1957 (in Russian).

[34] 34. **G. W. Swan and G. R. Fowles**. "Shock Wave Stability". *Phys. Fluids*, 18:28–397, 1975.

[35] 35. **G. R. Fowles**. "Stimulated and Spontaneous Emission of Acoustic Waves from Shock Waves". *Phys. Fluids*, 24:220–229, 1981.

[36] 36. **R. W. Griffiths and R. J. Sandeman and H. G. Hornung**. "The Stability of Shock Waves in Ionizing and Dissociating gases". *J.Phys.D: Appl. Phys.*, 9:1681–1693, 1976.

[37] 37. **S. A. Egorushkin and V. S. Uspenskii**. "Influence of Advanced Radiation on Stability of Strong Shock Waves in Gases with Arbitrary Equation of State". *Izvestiya AN SSSR, MZhG*, (3):125–133, 1990 (in Russian).

[38] 38. **Ya. I. Frenkel**. *Kinetic Theory of Liquid*. Dover, New York, 1955.

[39] 39. **A. G. Bashkirov**. "Nonequilibrium statistical thermodynamics of nucleation". *Journ. Chem. Physics*, 96(2):1484–1489, 1992.

[40] 40. **A. G. Bashkirov**. "Statistical Theory of Nucleation Kinetics". *Theor. and Math. Phys. (USSR)*, 3(2):497–504, 1970.

[41] 41. **A. G. Bashkirov and D. N. Zubarev**. "The Generalized Kramers- Fokker-Plank Equation". *Physica*, 48(2/3):137–146, 1970.

[42] 42. **J. Lothe and G. Pound**. "Reconsideration of Nucleation Theory". *Journ. Chem. Phys.*, 36:2080–2091, 1962.

[43] 43. **M. Sharaf and R. A. Dobbins**. "A Comparision of Measured Nucleation Rates with the Predictions of Sever Theories of Homogeneous Nucleation". *Journ. Chem. Phys.*, 77(3):1517–1526, 1982.

[44] 44. **A. G. Bashkirov**. "A reconsideration of nucleation theory". *Physics Letters*, A28(1):23–24, 1968.

[45] 45. **A. G. Bashkirov and A. G. Sutugin**. "On a Reconsideration of Classical Nucleation Theory". *Journ. Statist. Phys.*, 3(1):87–89, 1971.

[46] 46. **R. E. Prange**. "Tunneling from a Many–Particle Point of View". *Phys. Rev.*, 31:1083–1088, 1963.

[47] 47. **R. Fox and G. E. Uhlenbeck**. "Contribution to Nonequilibrium Thermodynamics. I. Theory of Hydrodynamic Fluctuations". *Phys. Fluids*, 13(8):1893–1902, 1970.

[48] 48. **N. N. Bogoliubov and N. M. Krylov**. On the Fokker–Planck equations. In *Bogoliubov N. N. Selected Works*, Volume 2, p. 5–76. Naukova Dumka, Kiev, 1970 (in Russian).

[49] 49. **J. Kirkwood**. "The Statistical Mechanical Theory of Transport Processes". *Journ. Chem. Physics*, 14:180–196, 1946.

[50] 50. **J. Lebowitz and E. Rubin**. "Dynamical Study of Brownian Motion". *Phys. Rev.*, 131:2381–2394, 1963.

[51] 51. **R. J. Cukier and J. M. Deutch**. "Microscopic Theory of Brownian Motion: The multiple-time-scale Point of View". *Phys. Review*, 177(1):240–244, 1969.

[52] 52. **M. Tokuyama and I. Oppenheim**. "Statistical–Mechanical Theory of Brownian Motion. Translational Motion in an equilibrium fluid". *Physica*, 94A:501–520, 1978.

[53] 53. **A. G. Bashkirov**. "Theory of Brownian Motion Due to Fluctuations of the Fluid ess Tensor". *Theor. and Math. Phys. (USSR)*, 44(1):95–102, 1980.

[54] 54. **R. Zwanzig**. "Ensemble Method in the Theory of Irreversibility". *Journ. Chem. Physics*, 33(5):1338–1341, 1960.

[55] 55. **D. ter Haar**. *Introduction to the Physics of Many-Particle Systems.* Interscience Publ., New York, 1958.

[56] 56. **L. D. Landau and E. M. Lifshitz**. *Fluid Mechanics.* Pergamon Press, New York, 1958.

[57] 57. **D. N. Zubarev and A. G. Bashkirov**. "Statistical Theory of Brownian Motion in a Moving Fluid in Presence of a Temperature Gradient". *Physica*, 39:334–340, 1968.

[58] 58. **A. G. Bashkirov**. "Statistical Theory of Thermodiffusion of Brownian Particles". *Theor. and Math. Phys. (USSR)*, 1(2):275–280, 1969.

[59] 59. **R. Mazo**. "On the Theory of Brownian Motion II. Non-uniform systems". *Journ. Statist. Physics*, 1(1):101–106, 1969.

[60] 60. **T. Munakata**. "A Kinetic Approach to the Friction Constant of a Brownian Particle". *Progr. Theor. Phys.*, 58(1):59–76, 1977.

[61] 61. **I. A. Michaels and I. Oppenheim**. "Long-Time Tails and Brownian Motion". *Physica*, 81A:221–240, 1975.

[62] 62. **G. Nicolis**. "On the Evaluation of the Thermal–Diffusion Coefficient of Heavy Particles Using a Theory of Brownian Motion in a Nonuniform Medium". *Journ. Chem. Physics*, 43(4):1110–1112, 1965.

[63] 63. **D. N. Zubarev and A. G. Bashkirov**. "Statistical Theory of Brownian Motion in a Temperature–Wise Inhomogeneous Moving Fluid". *Physics Letters*, 25a(3):201–203, 1967.

[64] 64. **J. Luttinger**. "Theory of Thermal Transport Coefficients". *Phys. Rev.*, 135A(6A):1505–1514, 1964.

[65] 65. **J. D. Ramshaw**. "Brownian Motion in a Flowing Fluid". *Phys. Fluids*, 22(9):1595–1601, 1979.

[66] 66. **J. M. Schurr**. "The Thermodynamic Driving Force in Mutual Diffusion of Hard Spheres". *Chem. Physics*, 65:217–223, 1982.

[67] 67. **M. M. Kops–Werkhoven and A. Vrij and H. N. W. Lekkerkerker**. "On the Relation Between Diffusion, Sedimentation and Friction". *Journ. Chem. Physics*, 78(5):2760–2763, 1983.

[68] 68. **P. Mazur**. “On the Motion of n Spheres in a Viscous fluid”. *Physica*, 110A(1):128–146, 1982.

[69] 69. **G. K. Batchelor**. “Sedimentation in a Dilute Dispersion of "Spheres”. *Journ. Fluid Mech.*, 52(2):245–268, 1972.

[70] 70. **E. J. Hinch**. “An Averaged–Equation Approach to Particle Interactions in a Fluid Suspension”. *Journ. Fluid. Mech.*, 83(4):695–720, 1977.

[71] 71. **R. M. Mazo**. “On the Theory of Brownian Motion III. Two–Body Distribution Function”. *Journ. Stat. Phys.*, 1:559–570, 1969.

[72] 72. **J. M. Deutch and I. Oppenheim**. “Molecular Theory of Brownian Motion for Several Particles”. *Journ. Chem. Phys.*, 54(8):3547–3555, 1971.

[73] 73. **T. J. Murphy and J. L. Aguerre**. “Brownian Motion of N Interacting Particles. I. Extension of Einsstein Diffusion Relation to the N–Particle Case”. *Journ. Chem. Phys.*, 57(5):2098–2104, 1972.

[74] 74. **J. L. Aguerre and T. J. Murphy**. “Brownian Motion of N Interacting Particles. II. Hydrodynamical evolution of the Diffusion Tensor Matrix”. *Journ. Chem. Phys.*, 59(4):1833–1840, 1972.

[75] 75. **B. Felderhof and R. Jones**. “Linear Response Theory of Sedimentation and Diffusion in a Suspension of Spherical Particles”. *Physica*, 119A:591–608, 1983.

[76] 76. **J. N. Roux**. “Brownian Particles at Different Time Scales–A New Derivation of the Smoluchowski Equation”. *Physica A*, 188(4):526–552, 1992.

[77] 77. **H. J. H. Clercx and P. P. J. M. Schram**. “Retarded Hydrodynamic Interactions in Suspensions”. *Physica A*, 174(2–3):325–354, 1991.

[78] **G. K. Batchelor**. “Brownian Diffusion of Particles with Hydrodynamic Interaction”. *Journ. Fluid Mech.*, 1:1–29, 1976.

[79] 79. **A. G. Bashkirov**. “Fluctuation Hydrodynamics of the Brownian Motion of a Particle in a Fixed Dispersed Layer”. *PMM Journ. Applied Math. and Mech. (USSR)*, 49(4):428–433, 1985.

[80] 80. **P. G. Saffman**. “On the Settling Speed of Free and Fixed Suspensions”. *Stud. Appl. Math.*, 52(2):115–127, 1973.

[81] 81. **I. D. Howells**. "Drag Due to the Motion of a Newtonian Fluid through a Sparse Random Array of Small Fixed Rigid Objects". *Journ. Fluid Mech.*, 64(3):449–475, 1974.

[82] 82. **D. Bedeaux and P.Mazur**. "Renormalization of the Diffusion Coefficient in a Fluctuating Fluid. III. Diffusion of Brownian Particle with Finite Size". *Physica*, 80A:189–202, 1975.

[83] 83. **J.—P. Carton and E. Dubois-Violette and J. Prost**. "Brownian Diffusion of a Small Particle in a Suspension. II. Hydrodynamic Effect in a Random Fixed Bed". *Physica*, 119A:307–316, 1983.

[84] 84. **D. L. Koch and J. F. Brady**. "Dispersion in Fixed Beds". *Journ. Fluid Mech.*, 154:399–427, 1985.

[85] 85. **A. G. Bashkirov and A. V. Orlov**. "Transverse Diffusion in a Granular Bed". *Theoret. Foundations of Chem. Engn.*, 25(5):540–546, 1991.

[86] 86. **Ch. Emiliani**. *Dictionary of the Physical Sciences.* Oxford University Press, New York, Oxford, 1987.

Index